RADIOMETRIC SYSTEM DESIGN

RADIOMETRIC SYSTEM DESIGN

CLAIR L. WYATT

Electrical Engineering Department
and Center for Space Engineering
Utah State University

MACMILLAN PUBLISHING COMPANY
A Division of Macmillan, Inc.
NEW YORK

Collier Macmillan Publishers
LONDON

Macmillan Publishing Company
866 Third Avenue, New York, NY 10022

Collier Macmillan Canada, Inc.

Printed in the United States of America

printing number
1 2 3 4 5 6 7 8 9 10

Library of Congress Cataloging-in-Publication Data

Wyatt, Clair L.
  Radiometric system design.

  Includes index.
  1. Electrooptical devices.   2. Optical data processing.
3. System design.   I. Title.
TA1750.W93   1987   621.36'7   87-7855
ISBN 0-02-948800-1

Preface

This book provides a system-oriented approach to radiometric design. It is written for the senior and first-year graduate student of engineering and/or science.

The scope of this book is limited to that of electro-optical systems for information processing and does not cover systems designed to utilize optical energy for work. The emphasis is upon systems wherein information is conveyed by means of an electro-optical beam such as in remote sensing, guidance and tracking, and fiber optic and laser communication systems with emphasis upon radiometric detection. Any system must be capable of detecting incident electromagnetic flux before any data processing can occur; thus, attention is also given to the optimization of systems for maximum signal-to-noise ratio.

Any electro-optical system for information processing can be modeled in terms of six basic subsystems: the source, intervening media, optical subsystem, focal plane, signal-conditioning electronics, and output and display.

The pedagogical approach is twofold: (1) to trace the signal flow from the source through the system and (2) to develop a radiometric performance equation in terms of the signal-to-noise ratio for optimization and sensitivity analysis.

The book is divided into two parts. Part I (Chaps. 1 through 9) considers that level of design in which system and subsystem figures of merit are defined and utilized to accomplish a radiometric feasibility study. The feasibility study includes source characterization, flux transfer, optical throughput, detection, noise, and uncertainty considerations to arrive at system performance objectives. The consideration of system and subsystem performance in terms of figures of merit provides an overview of important design criteria and should prepare a person to direct and monitor the development of a system without a knowledge of detailed subsystem design.

Part II discusses detailed design considerations of various configurations of some of the subsystems introduced in Part I. No single book could ever provide an all-inclusive coverage of the entire field of electro-optics, nor could any one author be an expert in all aspects of system and subsystem design. Part II necessarily reflects and is limited to the author's expertise resulting from some 25 years associated with the Center for Space Engineering of Utah State University.

Chapters 10 and 11 discuss practical aspects of blackbody radiation and optical media. This is accomplished by solving Planck's and Maxwell's equations. Chapters 12 through 18 provide detailed design information for some sensor subsystems. Chapter 19 covers calibration and error analysis. Emphasis is placed upon achieving state-of-the-art performance in terms of the system signal-to-noise ratio by considering the trade-offs necessary to achieve system specifications. Detailed design criteria are presented for several aspects of optical systems. These include basic optical configurations, modulation transfer function, and baffling; operation of thermal, photon noise limited, and multiplier phototube detectors; and design of electronic signal-conditioning subsystems, including low-noise preamplifiers and detector coupling.

Numerous example problems are provided in the text to illustrate numerical performance and dimensional analysis as applied to real-world problems. Student exercises are also provided for each topic.

It is not possible to mention all those who have contributed to the contents of this book. However, I am indebted to Dr. Doran J. Baker, Chairman of the Electrical Engineering Department of Utah State University; to Allan J. Steed, Director of the Center for Space Engineering; to students who took the classes; to members of the CSE staff; and to Tes Mace and Melinda Vance, who typed most of the manuscript. The research from which the original notes were written was supported by the Air Force Geophysics Laboratory, Bedford, Massachusetts. Drs. A. T. Stair, Jr., and R. E. Murphy in particular are acknowledged.

Contents

one

FIGURES OF MERIT AND FEASIBILITY STUDY

Part 1 considers the design level in which system and subsystem figures of merit are defined and utilized to accomplish a radiometric *feasibility study*. This study includes source characterization, flux transfer, optical through-put, detection, and noise considerations to arrive at *system performance objectives* based on subsystem figures of merit. The development of a *radiometric performance equation* provides for optimization and sensitivity analysis in terms of the system signal-to-noise ratio. This approach, using the generalized system model, provides an overview of important design criteria that apply to any electro-optical system designed for processing information.

chapter *1*

Electro-optical Systems

1.1 INTRODUCTION

Electro-optical engineering is an extremely diverse field dealing with many problems and solutions. The subject matter of this book is limited to systems for *information processing* and does not include systems designed to utilize optical energy for work. We are concerned primarily with information that is conveyed by means of a beam of *electro-optical* energy rather than electric charge as in more conventional communication systems.

This book presents a system-oriented approach: the first step is to define a generalized electro-optical system in which, by varying the nature of the subsystems and components, we can realize numerous important processing applications. The specific applications for which this generalized electro-optical system is described are remote sensing (including environmental and earth resources, laboratory spectroscopy, and guidance and tracking) and point-to-point communications.

Technical terms and definitions are introduced as needed. A summary of the important terms, symbols, and units used in the text is given in the glossary at the end of the book.

The field of electro-optical engineering results from the merger of two related fields: *electrical* engineering and the science of optics. *Electrical engineering* is concerned primarily with optimizing the generation, storage, transmission, control, and conversion of electric energy. In some cases, energy for doing work is the important quantity; in others, information is the important quantity, with energy providing the means for conveying that information. *Optics* is the science that deals with light—its creation and propagation, the changes it undergoes, and the effects it produces. The science of optics is not limited to those regions of the electromagnetic spectrum that produce sensory vision; many of the principles that apply to visible light apply equally well to the ultraviolet and infrared regions of the spectrum.

Electro-optical engineering might be defined as the engineering field that is primarily concerned with optimizing the generation, storage, transmission, control, and conversion of *optical energy*. In analogy with electrical engineering, in some cases optical energy for doing work is the important quantity, whereas in others information is the important quantity and optical energy provides the means for conveying it.

1.2   THE ELECTRO-OPTICAL SYSTEM

The basic building blocks of a generalized electro-optical system for remote sensing or for communications are given in Fig. 1.1. First there must be a source of radiation, shown at the left in Fig. 1.1. After having been modified by passing through an intervening medium and selectively collected by the optical subsystem, the radiation is incident upon a detector. The detector output (an electric signal) is amplified and filtered by the signal-conditioning electronic circuits. Finally, the output and display provides visual, graphic, or audio output (in communication systems) or control voltages and feedback (in guidance systems) and often includes inverse-transform (decoding) procedures in multiplexed systems.

All electro-optical systems designed for information processing utilize the basic building blocks shown in Fig. 1.1. These may be considered under the major headings of remote sensing and point-to-point communications.

1.2.1   Remote Sensing

Remote sensing deals with the *characterization* of a remote object through measurement of the radiation properties of the object. The attributes of an object cannot be measured directly by remote sensing of radiant flux but must be *inferred* from the sensor's response to the flux incident upon the entrance aperture. An object can be characterized using electro-optical systems in five nearly independent domains: *spatial* (position, size, and shape), *spectral* (distribution of energy as a function of wavelength or frequency), *temporal* (variation of flux with time), *polarization* (orientation

Figure 1.1   A functional flow diagram for a general electro-optical system.

of the electric field **E** vector), and *coherence* (phase of the elemental radiators).

Every object in the universe is emitting and absorbing radiation continually unless it is at absolute zero temperature. In addition, the object may be reflecting ambient or artificial radiation. It is possible to characterize a remote object using emitted, absorbed, or reflected radiation.

Remote sensing takes many forms. For example, the location of a remote object may be determined using a simple spatial scheme, or a more complex imaging system can provide a characterization of an entire scene. The temperature, physical process, or energy levels associated with a plasma can be determined from its spectral signature. The chemical composition of a substance can be determined from its absorptance or emission spectra. The properties of a contaminative film on the surface of a solid can be determined using polarization techniques.[1]

Coherence (or partial coherence) has significant metrological implications that must often be taken into account in the design of high-resolution optical systems.[2]

Discrimination is a measure of the ability to detect a signal that is buried in noise.[3]

A target can be characterized using the spatial, spectral, temporal, polarization, and coherence properties of radiation, provided the radiation contains sufficient information. Statistical decision theory is often used to determine the presence or absence of an object in the scene.[4] A more general approach to the problem of scene characterization is obtained using pattern recognition theory[5] by means of which the scene can be classified into a number of useful categories.

Passive and active systems are defined according to whether the radiation emanating from the remote object results from self-emission or reflection of radiant energy from a natural source such as the sun, or from artificial source, such as a laser. Examples of *passive* systems are missile guidance systems that detect the heat radiating from a target such as a jet engine or engine rocket exhaust plume and the earth resources satellite (LandSat), which makes use of a multispectral scanner that detects reflected solar light. An example of an *active* system is a laser guidance system, which seeks an object that is illuminated by a directed laser beam.

1.2.2 Point-to-Point Communications

Communication systems are utilized to *transfer information* from point to point. There are two general categories of information: scene representation (visual) and data transfer. Visual data communication systems include photography, television, and radar. Lasers provide the primary optical means of data transfer. Lasers, utilized as transmitters, exhibit great potential for high-data-rate communications applications.[6] Their narrow transmit beam and narrow receiver field of view also provide the desirable characteristics of inherent privacy and jam resistance. Additional advantages of laser systems are their immunity to nuclear detonation ionization effects, survivability; covertness; and small size, weight, and power requirements. Laser systems have been proposed for satellite-to-ground communications for operation under favorable atmospheric conditions.

The potentially high information-carrying capacity of optical waveguides (fiber optics), along with their convenient size and low weight, their immunity from radio-frequency and electromagnetic interference, and above all their very low cost potential, are reasons for further development and application of ground-to-ground communication systems.

1.3 THE ELECTROMAGNETIC SPECTRUM

Energy transferred from one point to another in a beam of radiant energy can be considered either as being propagated by wave motion or as a stream of photon particles. Figure 1.2 identifies wavelength intervals within which a common body of *experimental techniques* exist. These wavelength regions are rather loosely defined and overlap.

The basic units of the *International System* (Système International, or SI) consisting of the meter (m), kilogram (kg), and second (s) are used throughout this text (see Appendix A). However, an exception is made in the case of wavelength and wavenumber as used to designate the spectrum; the centimeter (cm) is used for distance rather than the meter.

The Greek letter lambda (λ) is used to designate *wavelength*, which is defined as the distance between two consecutive points in a wave that have the same phase. The most common unit for wavelength is the *micrometer* (μm), where μ (Greek letter mu) is a prefix that represents 10^{-6} (see Appendix B for a list of prefixes commonly used with the SI system). Alternative units for wavelength are the *nanometer* (nm), where n represents 10^{-9}, and the *angstrom* (Å), which is defined as 10^{-10} m. The symbol used to represent *wavenumber* is ν, where wavenumber is defined as the number of wavelengths that would, at an instant, occupy a space 1 cm in length. The units for wavenumber must therefore be reciprocal centimeters:

$$\bar{\nu} = \frac{\nu}{c} = \frac{1}{\lambda} \quad [\text{cm}^{-1}] \quad (1.1)$$

where c is the velocity of light and ν is the frequency in hertz (1 Hz = 1 cycle per second).

The spectral ranges considered in this book are categorized on the basis of distinct detection techniques as shown in Table 1.1. Electro-optical systems are wavelength-dependent because of the nature of optical materials.

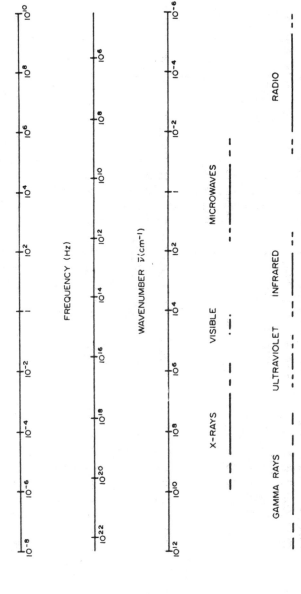

Figure 1.2 The electromagnetic spectrum. (Note: The scale is logarithmic.) (From C. L. Wyatt, *Radiometric Calibration: Theory and Methods*, Academic Press, New York, 1978, p. 8. Used with permission.)

Table 1.1 REGIONS OF THE OPTICAL SPECTRUM

Ultraviolet (UV)	1,800–4,500 Å
Visible (VIS)	4,500–7,500 Å
Near infrared (NIR)	7,500–10,000 Å
Short-wave infrared (SWIR)	1–5 μm
Intermediate-wavelength infrared (IWIR)	5–25 μm
Long-wavelength infrared (LWIR)	25–1000 μm

1.4 SPECIFIC EXAMPLES OF ELECTRO-OPTICAL SYSTEMS

Electromagnetic radiation in the optical portion of the spectrum interacts with materials in complex ways. This is fortunate, because it is the fact that optical radiation is modified by reflection off or transmission through various substances that provides the signatures by which remote objects or processes can be characterized. However, the material's optical characteristics also result in numerous limitations for electro-optical systems, and the design is rendered more complex.

It was pointed out in the introduction to this chapter that information processing using optical energy transfer is analogous to that of more conventional electric circuits using electric charge transfer. However, in the electro-optical system the "circuits" must be considered as "distributed" rather than "lumped" as in simple electronic circuits. Optical circuits have no equivalent to the zero-impedance wire.

The following subsections illustrate specific examples of electro-optical systems that can be represented by the functional-flow diagram of Fig. 1.1.

1.4.1 Remote-Sensing Systems

A system designed to collect radiant energy from a remote scene and classify the elements of the scene into several categories can be represented by Fig. 1.1. Reflected and emitted radiant energy from the scene objects is incident upon the optics after having passed through and been modified by the intervening medium (atmosphere). Unwanted radiation, coming directly from the source and the background, also falls on the optics.

The optical and detector subsections are designed to determine the spatial (field-of-view) and spectral (wavelength) response characteristics of the sensor to reduce its sensitivity to off-axis sources and to limit its response to those portions of the reflected or emitted spectrum that are most characteristic of the objects of interest in the scene. The temporal, polarization, and coherence properties of the optical and detector subsections can also be designed to utilize whatever information the radiation may contain in those domains.

The signal-conditioning system is designed to optimize the signal-to-noise ratio and to determine the system's temporal response.

The output and display subsystem may be capable of generating a two-dimensional raster display of the scene processed to enhance objects of interest, or it may classify the scene into several categories or objects of interest, or it may simply make a decision as to the presence or absence of a particular object within the scene. The display subsystem may perform any one of various inverse-transform procedures to unscramble the multiplexed data.

Active systems utilize artificial sources (laser beams, for example) to illuminate the scene, so the unique characteristics of the source are incorporated into the design.

1.4.2 Laboratory Spectrometer

Spectroscopic analysis is a specialized application of remote sensing that can also be represented by Fig. 1.1. In this case an artificial source provides a smooth and continuously varying distribution of energy as a function of wavelength. This radiant energy is reflected off a sample of the substance to be analyzed in reflectance spectroscopy or is transmitted through the sample to be analyzed in absorption spectroscopy. The optics and detector subsystems function to determine the field of view and spectral response as described in Sec. 1.4.1. The polarization properties of the source, sample, and sensor create important metrological problems that must be considered. The electronics and display subsystems can provide a graphic representation of the spectra or digital data output for computer graphics processing.

1.4.3 Missile Guidance System

Figure 1.1 can also represent a missile guidance system. In the case of an active system, the source may be a pulsed laser directed at a target. The optics and detector subsystems search and lock on the pulse-coded laser radiation reflected off the target. The electronics and output subsystems include servomechanisms to scan the optics and direct the missile toward the target.

For a passive missile guidance system, the target may be a self-emitting jet engine of an enemy aircraft. The optics and detector subsystems must be designed to discriminate between the target and competing sources of radiant energy such as the earth, sun, or sunlit clouds. The atmosphere (medium) constitutes a degrading feature of the system. Clouds, fog, and haze may limit the system's application to those occasions or locations (high altitude) where conditions are acceptable.

1.4.4 Communications Link

Figure 1.1 can represent a communications link. In this case, the source might be a laser encoded with data, and the medium could be the atmosphere or an optical waveguide (fiber optics). The optics and detector collect the beam or couple the radiant energy from the fiber to the detector. The electronics and output subsystems correspond to the decoding and switching functions of complex communication systems.

1.5 SYSTEM NOMENCLATURE

There are a number of terms used to describe electro-optical sensors in the literature. These terms are not used consistently, but they do convey some indication of the type of measurement intended for a particular instrument. They are generally composed of combinations of the prefixes and suffixes listed in Table 1.2.

Radio refers to electromagnetic radiation in general. *Photo* refers to visible radiation. *Spectro* refers to the division of the radiation into components. *Meter* implies a measurement is indicated, but the method of presentation is not specified. *Graph* implies that the measurement is recorded in some graphic form. *Scope* implies that the radiation can be viewed by the eye through the device. Thus, a *radiometer* is a sensor used to measure radiation. However, the term generally implies that an absolute measurement of radiant flux is made over a specific wavelength interval.

Additional adjectives may be used to specify the spectral interval that the particular instrument measures, such as *infrared radiometer*, *ultraviolet radiometer*, or *microwave radiometer*.

A *photometer* is an instrument used in the visible range, and it is implied that the response is adjusted to give measurements in terms of the visual effect. However, any sensor that uses a multiplier phototube as the detector is commonly referred to as a photometer regardless of the spectral range.

In a similar manner, *spectrometer* is a general term for any device used to measure radiation at selected wavelengths. The spectrometer measures the distribution of radiant flux as a continuous function of wavelength throughout a limited region of the electromagnetic spectrum that is defined

**Table 1.2 PREFIXES AND SUFFIXES
USED TO DESIGNATE SENSORS**

Prefix	Suffix
Radio	meter
Photo	graph
Spectro	scope

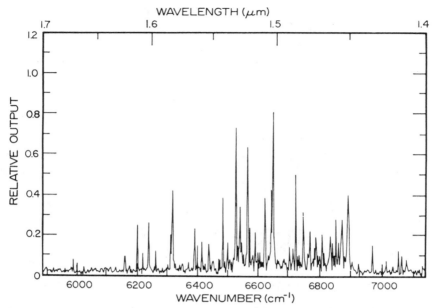

Figure 1.3 The airglow spectrum of the hydroxyl obtained with the Utah State University field-widened interferometer-spectrometer. (From C. L. Wyatt, *Radiometric Calibration: Theory and Methods*, Academic Press, New York, 1978, p. 10. Used with permission.)

as the *free spectral range*. This distribution is known as the *power spectral density function* or simply the *spectrum*. Figure 1.3 shows the airglow spectrum of the hydroxyl radical obtained with a field-widened Michelson interferometer spectrometer.

The *spectrograph* is a spectrometer that produces a graphically recorded spectrum as an inherent part of its function. Usually the term "spectrograph" is used to imply that the recording is done on photographic film placed in the image plane. Such a recording is called a *spectrogram*.

A *recording spectrometer* is a spectrometer that produces a graphic plot of power as a function of wavelength rather than as a photograph of the spectrum. A *spectroscope* is a spectrometer that provides for direct observation with the eye.

A *spectro-radiometer* is a spectrometer that has been calibrated for absolute measurements of a spectrum. Such an instrument functions in such a way that it is roughly equivalent to n radiometers, where n is the number of resolution elements within the free spectral range of the spectrometer. An absolute calibration of a spectro-radiometer is very difficult to obtain because of the great complexity of that type of instrument. Often a spectrometer is used only to measure the relative power spectral density while an associated radiometer obtains data to provide an absolute scale on the relative power spectral density function obtained with the spectrometer.

Radiant energy can be broken out into its component wavelengths by dispersion techniques or by interferometric techniques. A *dispersion spectrometer* is a spectrometer that employs either a prism or a grating to separate the energy into its spectral components. An *interferometer spectrometer* utilizes interference phenomena to separate the energy into its spectral components.

Sequential and multiplex spectrometers differ in the way the data are acquired: A *sequential* spectrometer obtains an independent measurement of each component of the spectrum sequentially—one after the other—over time. A *multiplex* spectrometer measures all the spectra at the same time; each component of each spectrum is encoded. Inverse-transform techniques are required to extract the spectral information from the encoded data. The Michelson interferometer spectrometer is an example of a multiplex spectrometer that uses Fourier transform techniques. The Hadamard spectrometer utilizes other multiplexing techniques where the inverse transform constitutes a form of co-adding.

1.6   SYSTEM CALIBRATION

In order to determine the temporal, spatial, spectral, polarization, or coherent properties of a beam of electromagnetic radiation emanating from an object, it is necessary to calibrate the sensor system in those same domains. Success in defining object attributes using remote-sensing techniques or in conveying information on electro-optical communication systems requires that the sensor response in those domains be defined such that the sensor attributes *contribute* to object characterization.

System calibration can be visualized as the quality control aspect of system design and testing. The system performance and conformity to specifications can be determined only by calibration. The necessity to interpret field data imposes additional requirements upon the calibration procedures and reporting techniques.

EXERCISES

1. Name the subsystems in the generalized electro-optical system.
2. What is discrimination in remote sensing?
3. What is pattern recognition in remote sensing?
4. Give the general nomenclature for a sensor that yields:
   (a) Total integrated measurements
   (b) Power spectral density function
5. Define the optical spectrum.
6. Describe the function of a spectrograph sensor.
7. Give a possible name for a sensor that uses a multiplier phototube to measure the total energy in an ultraviolet band.

8. Refer to Appendix B, and complete the following table with the missing equivalent measures.

Wavelength, μm	Wavelength, nm	Wavelength, Å	Wavenumber, cm⁻¹	Frequency, Hz
1.0	—	—	—	—
10.0	—	—	—	—
—	200	—	—	—
—	750	—	—	—
—	—	5000	—	—
—	—	10000	—	—
—	—	—	1200	—
—	—	—	6800	—
—	—	—	—	10^{14}

9. What are the two major functions of calibration of electro-optical sensors designed and fabricated for instrumentation applications?

REFERENCES

1. K. Vedam, "Applications of Polarized Light in Materials Research," *Proc. Soc. Photo-Opt. Instrum. Eng.—Polarized Light*, **88**, 78–83 (1976).
2. G. O. Reynolds and J. B. DeVelis, "Review of Optical Coherence Effects in Instrument Design," *Proc. Soc. Photo-Opt. Instrum. Eng.—Applications of Optical Coherence*, **194**, 2–33 (1979).
3. W. W. Harman, *Principles of the Statistical Theory of Communication*, McGraw-Hill, New York, 1963, pp. 227–248.
4. W. B. Davenport, Jr., and W. L. Root, *An Introduction to the Theory of Random Signals and Noise*, McGraw-Hill, New York, 1958, pp. 312, 364.
5. K. S. Fu, "Pattern Recognition in Remote Sensing of the Earth's Resources," *IEEE Trans.—Geoscience Electronics*, **6e-14**(1), 10–18 (1976).
6. J. R. Roland and C. E. Whited, "Air Force Space Laser Communications," *Proc. Soc. Photo-Opt. Instrum. Eng.—Laser and Fiber Optics Communications*, **150**, 2–7 (1978).

chapter *2*

Electro-optical Design

2.1 INTRODUCTION

While pure science is concerned with the search for basic knowledge, engineering is the application of knowledge to serve human needs. Thus, *engineering* is defined as the application of science and mathematics by which properties of matter and sources of energy in nature are made useful to people in structures, machines, products, systems, and processes.[1] Engineers do the things required to serve the needs of the people; they apply scientific theory and mathematical techniques and in so doing provide for the material needs and well-being of the world's inhabitants. However, the needs and the tools have changed through the years.

Engineering design is an iterative decision-making process intended to produce plans by which various material resources are organized, preferably in an optimal manner, into structures, machines, and products to meet human needs.[2]

A great many books have been written upon the subject of the design process.[3] Some authors focus upon techniques of optimization and have as one objective to optimize the entire process of design. Others stress the process of optimizing components of design. Analysis of the design process can become very complex. Figure 2.1 is one of the many possible representations of the design process as a functional flow diagram.

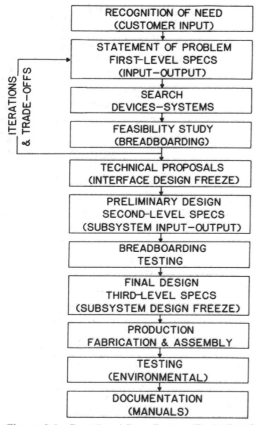

Figure 2.1 Functional flow diagram illustrating the design process.

2.1.1 Trade-offs

At every step in the design process, the engineer is confronted with conflicting needs that require trade-offs. Many desirable attributes of a system are mutually exclusive; that is, one can be obtained only at the expense of another. For example, the noise-equivalent power of a sensitive radiometer is proportional to the square root of the noise bandwidth or inversely proportional to the square root of the rise time. One cannot have both great sensitivity and fast response. Other trade-offs relate to system capabilities versus cost and complexity of operation, or to cost versus reliability, for example.

2.1.2 Iteration

Iteration occurs to some extent at all levels. It arises from the fact that hindsight is better than foresight. The engineer can never anticipate all the ramifications of the problem. However, experience helps him or her become more proficient.

It would be nice if the design process proceeded without deviation through each phase in order. This rarely happens. The engineer may find it necessary to back up and start something all over again. This can take place as a direct result of the fact that the engineer's understanding of the very nature of the problem changes as she or he obtains greater insight into it through experience in working on it. These changes in the concept of the problem may require new interaction with the customer.

Iteration is one of the special features that distinguishes the design process from analysis or research. In analysis, the starting point is given and the solution is determined by the analytic method. In design, the end point is given and a starting point must be obtained by synthesis. However, during the process of design, one attempts to work from an assumed starting point to the required end. As errors in the direction are detected, backing up to a new starting point is frequently required—this is *iteration*.

2.1.3 Optimum Design

Of several possible approaches available, the engineer, through the analysis phase, must determine which one is optimum. A relatively small class of problems exist for which *synthesis* of *optimum* systems is possible.[4] The optimum synthesis process has the advantage that we can be assured that it is impossible to construct a better device, subject to the given constraints. In other words, knowledge of the theoretically optimum system makes it unnecessary to consider a large number of possible modifications, just as the law of conservation of energy makes it superfluous to analyze different ways of constructing perpetual motion machines.

There are also some special cases in which an approach can be selected through a process of optimization. *Optimization* is a process in which various parameters are adjusted to maximize a desired result. However, it is necessary to establish the importance of various criteria by value judgments within given constraints. In this case, the engineer must decide what is most important among several features: cost, size, weight, reliability, etc. Even the "given constraints" must be chosen by value judgment.

The meaning of "optimum" can vary greatly depending upon the application and extenuating circumstances. *Minimum production cost and ease of maintenance* are primary considerations in the design of competitive consumer products. *Maximum reliability* and *minimum size* are requirements for military and space applications.

A primary consideration in electro-optical sensor design is the ability of the sensor to detect faint signals. Thus, one criterion for an optimum design is that the sensitivity is limited by fundamental effects that are inherent in the radiant source or the electro-optical detector and are therefore beyond the control of the designer. This is the aspect of electro-optical system design that is concerned with the electrical interface to the electro-optical detector. It includes considerations of low-noise preamplifiers, signal-conditioning amplifiers, phase-sensitive detection (dc restoration), control systems, and computer algorithms.

2.2 SYSTEM SPECIFICATIONS AND TOLERANCES

There are usually at least four levels of specifications that might be identified in electro-optical system design. First-level specifications deal with the user interface—the relationship between the human being and the system visualized as a black box.

The first effort consists of expressing the human need as a set of technical specifications. This requires defining the first-level specifications and tolerances, which, as illustrated in Fig. 2.1, leads to both iteration and trade-offs. The engineer must help the customer to define *performance objectives*. These objectives must not go beyond the possible or contain requirements that unnecessarily drive up the cost. Any statement of the technical specification must be justified in terms of the *system performance goals*.

A complex system is simplified by breaking it down into subsystems as illustrated in Fig. 2.2, which is a functional block diagram for a radiometer that utilizes phase-sensitive detection. Functional block diagrams are used to define subsystem specifications. Second-level specifications deal with subsystem definition and interface. Third-level specifications define the individual mechanical, optical, and electrical subsystem schematics or layouts. Finally, fourth-level specifications deal with the individual devices used in each subsystem. Very complex systems may require more than four levels of definition and specification.

Unfortunately, even the first-level specifications cannot be agreed upon without consideration of the individual devices to be used. For this

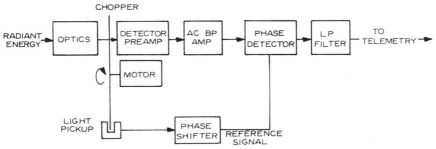

Figure 2.2 Illustration of a functional block diagram of a radiometer that utilizes phase-sensitive detection.

reason, a *feasibility study* is also required in the development of first-level specifications. Such a study may include some testing of hardware or the extrapolation of the design of existing systems to meet the new requirements.

The feasibility study is usually performed by highly competent engineers who possess advanced academic degrees and have had considerable experience in this field of work. A favorable feasibility study (if correctly and thoroughly carried out) guarantees that the customer will accept delivery of the system upon completion.

The preliminary design is accomplished by engineers who are familiar with similar designs that have proven successful. The conservative approach is to base the design upon existing systems for which relatively small changes (improvements or modifications) are needed to meet the new requirements. For this reason, companies tend to restrict their work to areas in which they have a history of success and to build upon that basis as a specialty.

Occasionally, satisfying the design specifications may require that systems be developed that extend far beyond the present experience level. The successful development of such systems may represent real advancement in the state of the art, in which case the specifications may be stated as "design goals." The customer and the design firm share the burden of the uncertainty of the outcome. The customer agrees to accept the system provided the engineering firm has made a reasonable effort to achieve the goals regardless of how far the performance may fall short of those goals. The success of a company depends upon the ability of its engineers to innovate unique approaches to problems.

The final production design requires engineers who have a good background in production and manufacturing processes.

A major characteristic of engineering related to the tradition of both science and engineering is the individual's ability to both *abstract* and *predict*. This provides the power to produce a design or plan (before the fact or existence of the system) and the confidence that in the end the system will function according to that plan.

2.3 SUBSYSTEM SPECIFICATIONS

The following description of possible subsystem specifications is based on the six basic building blocks of the generalized electro-optical system outlined in Sec. 1.2 and Fig. 1.1.

The *source* may be considered part of an electro-optical system even when it is part of the environment and beyond control. Its radiation properties, natural or artificial, as modified by the intervening medium, determine the required detector characteristics. However, the detector and source must be matched, and available detectors may dictate the type of source to be used in some applications. The source specifications may include but are not limited to the following:

1. Size, shape, range, movement
2. Temperature
3. Emissivity
4. Radiant properties:*
   Spectral sterance [radiance] L, in W cm^{-2} sr^{-1} μm^{-1}
   Spectral areance [exitance] E, in W cm^{-2} μm^{-1}
   Spectral pointance [intensity] I, in W sr^{-1} μm^{-1}
5. Temporal characteristics

The *intervening medium* modifies and in some cases severely limits the amount of radiant energy reaching the detector. The medium specifications may include but are not limited to the following:

1. Spectral emission, absorption, and scattering
2. Path length
3. Spectral transmittance

The *optical subsystem* determines the spatial, spectral, and polarization properties of the sensor. The spatial specifications for an electro-optical system may include but are not limited to the following:

1. The equivalent ideal *field of view* (expressed in terms of the solid angle or the linear angle for symmetrical systems)
2. Spatial scanning mode and range
3. The out-of-field or off-axis rejection

The spectral specifications for an electro-optical system may include but are not limited to the following:

1. Free spectral range
2. Spectral resolution
3. Out-of-band rejection

The polarization specifications for an electro-optical system may include but are not limited to the following:

1. The degree of polarization sensitivity
2. Polarization parameters (Stokes)

The *detector subsystem* is the heart of the electro-optical system, because detector performance is so materials-dependent and because it is the detector that transforms the optical radiation into an electric signal that can be amplified and processed. The detector subsystem (or what is often

*See Sec. 3.3 for a discussion of these terms.

referred to as the "focal plane") specifications may include but are not limited to the following:

1. Spectral region of response
2. Absolute responsivity
3. Noise equivalent power, responsivity, or detectivity
4. Operating temperature
5. Radiant background
6. Frequency response
7. Number of detector elements

The *signal-conditioning subsystem* often provides for low-noise impedance matching of the detector to the amplifier, current-to-voltage conversion (when required), gain, phase-sensitive detection (for chopped systems), and filtering. Thus, the signal-conditioning subsystem specifications may include but are not limited to the following:

1. Low-noise detector impedance matching preamplifier
2. Gain
3. Bandpass and/or lowpass filtering
4. Phase-sensitive detection

The *output and display subsystem* could consist of a meter or a form of graphical, visual, or oral presentation; it could include inverse transforms to unscramble multiplexed data; guidance control systems to direct a missile toward a target; or telemetry channels. Thus, the output and display subsystem specifications may include but are not limited to the following:

1. Output voltage range and source impedance
2. Inverse-transform algorithms
3. Control systems
4. Bit rates
5. Graphic display
6. Data storage media
7. Telemetry systems

Each of the above specifications parameters is defined in the chapters to follow.

2.4 SUBSYSTEM TOLERANCES

Engineering tolerance control is a problem because few engineers have had formal training in tolerance analysis and neglect has resulted in the lack of standardized techniques.[5]

The problem is further compounded by the fact that there are three separate groups in any manufacturing company who are constantly concerned with tolerances and tolerance buildup: Design Engineering, which is concerned with product developmental research; Manufacturing Engineering, which is concerned with the actual manufacturing process and assigns appropriate tolerances to tool, gauge, and shop operations; and Quality Control Engineering, which has the responsibility of measuring and evaluating the whole process. These three groups do not always use the same techniques or have the same orientation for analyzing tolerance relationships.

Assembly tolerance buildup is an arithmetic summation of the tolerances of each part in the assembly and gives the maximum possible range for a given dimension. Statistical tolerancing, as opposed to arithmetic tolerancing, is based upon the idea that the tolerances of individual parts have a statistical distribution and do not therefore combine arithmetically.

The predicted statistical assembly tolerances usually depend upon the assumption of natural (gaussian) distribution and certain empirical "design factors" and are stated in terms of "probable maximum variation in overall tolerance."[6]

A major difficulty in electro-optical design is that of determining the tolerances of one subsystem (e.g., optical) in time for the design of another subsystem (e.g., structural).

2.5   ELECTRO-OPTICAL SYSTEM DESIGN

The basic building blocks of an electro-optical system were presented in Fig. 1.1. It was pointed out that optical radiation is modified by reflection off or transmission through various media. Figure 2.3 illustrates the prob-

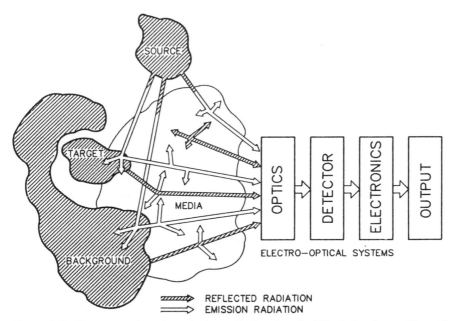

Figure 2.3 Representation of an electro-optical system that illustrates the problem of competing or interfering sources.

lem of radiation transfer and of competing or interfering sources that must always be considered in electro-optical design.[7] A target is shown embedded in a background and in an intervening medium and illuminated by a source. The target, the background, and the medium are represented as sources of reflected, scattered, and emitted radiant energy, and some of this energy reaches the optics. The design of an electro-optical system for any application must always address the problems of competing scattered and background radiation and the modifying effects of an intervening medium upon the radiation emanating from the object of interest.

Selectivity is achieved in an electro-optical system through careful consideration of the system performance objectives. For example, the optical subsection (see Fig. 2.3) can be designed so that the system responds only to energy that has its origin in a specific geometrical region. Energy falling upon the collecting aperture from anywhere outside the desired region is rendered ineffective in stimulating a response in the sensor. This example has reference to the *spatial domain* (see Sec. 1.2.1) by which an object embedded in a background, as in Fig. 2.3, can be characterized. Another term used in describing the spatial response of a sensor is *field of view*. The sensor field of view is described in terms of a linear half-angle for a circularly symmetric field of view (like that of a binocular) and in terms of a *solid angle* that is a measure of the three-dimensional cone associated with the boundary of the field of view.

The design of the sensor field of view must take into account system performance objectives, as described above, and include factors such as uniformity of response within the specified region, steepness of the falloff of response within the transition region as illustrated in Fig 2.4, and the off-axis response in the wings.

These aspects of a system field of view vary greatly in difficulty and complexity of design implementation and calibration. It is important that the system performance objectives be expressed in terms of system specifications for each. The degree of off-axis rejection required depends upon the nature of the background in which the object is embedded and the impact that it may have upon system performance objectives. The costs of design, implementation, and calibration are also related to the degree of off-axis rejection required. The designer must be alert to the possibility that certain specifications that may be extremely costly to implement may not be related to system performance objectives and may therefore represent an unnecessary cost. On the other hand, the designer must interact with the customer to be certain that such considerations are not overlooked altogether.

The foregoing discussion with respect to the spatial domain, or field of view, applies equally well to other aspects of the system. For example, the optics also determine the response in the *spectral domain*. Here the system can be designed to respond only to energy that has its origin in a specific wavelength region. Energy falling upon the collecting aperture from anywhere outside the desired spectral region must be rendered ineffective in stimulating a response in the sensor. The design of the sensor must consider

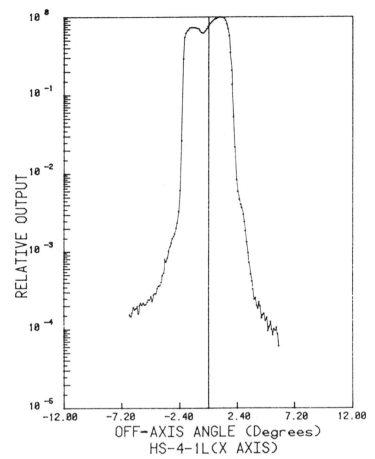

Figure 2.4 Illustration of steepness of the falloff of the point-source response, in the transition zone, of a sensor field of view.

the uniformity within the spectral domain, steepness of the falloff in the transition region, and out-of-band response in the wings.

　　Similar consideration must be given to the temporal, polarization, and coherence domains and to numerous factors such as environment, dynamic range, signal-processing techniques, and decoding of multiplexed signals. Each of these parameters must be considered in terms of system performance objectives.

2.5.1　Feasibility Study

The objective of the feasibility study is to guarantee that the first-level specifications can be achieved. The reputation of the engineer and his or her company is at stake. The feasibility study must (1) identify all possible alternatives that appear to satisfy the specifications; (2) analyze each in terms of performance, ease of implementation, and cost; and (3) select the preferred approach.[8]

Product and market planning are required as an extension to the feasibility study, particularly if it is a speculative venture rather than the response to a request for quote.

The trade-offs and iterations required to accomplish the feasibility study are illustrated in Fig. 2.5. Each alternative design must be evaluated. A *radiometric analysis* is the starting point of such a study. It requires selection of a detector of radiant energy and the associated optical and electrical parameters that will satisfy the specifications regarding source, radiation transfer through the intervening medium, system sensitivity, spectral response region, and response time. Stated simply: Information transfer utilizing a radiant beam requires that radiant energy be detected with an adequate signal-to-noise ratio.

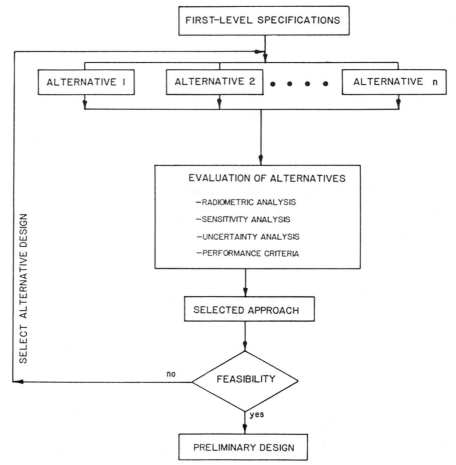

Figure 2.5 Feasibility study trade-offs and iteration as a functional flow diagram.

The selection of a specific detector, based upon a given figure of merit, requires that certain environmental conditions be attained, and particular background requirements may even dictate cryogenic cooling. The detector noise characteristics may require that signal-coding techniques be used; this may dictate that choppers or scanners be employed. Each of these and other considerations must be evaluated in the feasibility study for each candidate detector.

The source can be described in terms of the power (energy rate) or photon rate and various normalizing terms that include the geometrical characteristics of the radiation transfer. The most general source figure of merit is the sterance [radiance]. The medium can be described in terms of the attenuation, which is usually a function of wavelength and path. In some cases there is emission or scattering of radiation from out-of-field sources into the field of view.

Detector performance can be described in terms of various *figures of merit* (such as the *responsivity*, *noise equivalent power*, or *detectivity*). The *throughput* provides a figure of merit for the optical subsystem that is more or less independent of the specific optical design. In addition, the optical subsystem provides for spectral, spatial, and polarization selectivity. The electronics can be modeled in terms of the *gain* and *electrical noise bandwidth*. From these relatively simple parameters the responsivity of a system can be evaluated for a given spectral region and integration time.

The subsystem figures of merit can be combined to create overall system performance models. For example, an important parameter is the noise equivalent flux (NEF), which can be interpreted three ways, to yield (1) the smallest detectable change in the average radiant flux that can be detected, (2) the smallest average value of the absolute radiant flux that can be detected, and (3) the signal-to-noise ratio for a given average flux level.

The *sensor uncertainty* pertains primarily to metrological instrumentation systems. It is obtained in the system calibration and depends upon the sensor precision (repeatability) and the standard source uncertainty.

Additional system parameters can also be modeled, using suitable subsystem figures of merit, such as spectral resolving power, modulation transfer function (contrast) in imaging systems, spatial resolution in radiometric systems, and error rates in pattern recognition systems.

The ordering of evaluation criteria depends upon system performance objectives and the level and complexity of analysis. For optical systems the first-order *performance criterion* is often (as stated before) adequate signal-to-noise ratio for subsystems designed for information processing. Lower-order performance criteria probably arise from consideration of the quality of the data, which is related to spatial and spectral purity.[7]

Part 1 of this book deals with that level of design in which a radiometric feasibility study can be accomplished by making use of system and subsystem figures of merit. An objective of the feasibility study is to arrive at system performance objectives based upon appropriate figures of merit without recourse to detailed design.

Such a development does not usually require a knowledge of physical processes. For example, a source can be characterized in terms of certain geometrical figures of merit without regard to emission mechanisms, fundamental processes, etc.

The development in this text provides not only a rationale for a feasibility study, but also an overview of an electro-optical system design without cluttering up the picture with too much detail.

2.5.2 Preliminary Design

The preliminary design is undertaken after the first-level specifications have been established, the design freeze accepted, and the project funded. In the preliminary design, each of the subsystems is defined. The interface between subsystems must also be defined at this point. The interface definition is very important, as different engineers are frequently assigned to each subsystem. Often, a particular subsystem design cannot be started until the interfacing subsystem tolerances have been specified. It is important that subsystems be compatible when they are to be interfaced. Frequently, the more serious system problems result from failure to keep interface and tolerance requirements in mind during subsystem design.

2.5.3 Final Design

The final design results in a subsystem design freeze. It includes the detailed design and testing of subsystems.

The final design of the source is closely related to the medium and the detector. The source may be natural or artificial. In active communication systems the source must be chosen for its capability to be modulated at high frequencies. It may be chosen for its spectral characteristics to match a suitable detector.

The medium may be the atmosphere, an unknown sample to be analyzed, or a fiber optic. The final design will be based upon its transmittance, absorptance, and scattering properties.

Optical schematics, ray trace studies, and detailed mechanical design and component specifications are the result of the final design of the optical subsection. It includes those designs and components that determine the spatial and spectral characteristics of system performance. The design must also interface with the detector (which is referred to as the focal plane) both optically and environmentally. Not only does the optical subsystem focus the selectively collected radiant energy on the detector, but it may also share the same background and temperature environment required by the detector. This might, for example, include cryogenic cooling where packaging of the sensor optical head requires integration of the optics, detector, and preamplifier into the same unit.

The detector subsection design must take into account the detector and its interface with the signal-conditioning electronics. This includes the required temporal response rate and preamplifier noise and impedance levels.

The final design of the signal-conditioning subsystem results in component specification, electrical schematics, and mechanical layouts. This subsystem is frequently packaged separately from the optical head and is interconnected by means of appropriate electrical cables and connectors.

The output subsystem may take on any one of many forms depending upon the system performance objectives. It often includes inverse-transform procedures, servocontrol systems, or display devices. It may utilize real-time data processing or recording equipment and subsequent computer analysis.

2.5.4 Production and Testing

Final production, assembly, and testing usually uncover numerous problems resulting in "back to the drawing board" crises. This is normal, because no one can anticipate every possible problem. The design procedure includes iteration at every level. Often it is very difficult to estimate the amount of time required to complete production and testing.

An engineering calibration should be considered as the final phase of sensor fabrication at which time an engineering firm is considered to have established conformity with design specifications. Even at this point in the design process, it is sometimes necessary to go back to the customer and negotiate changes in the specifications that may result from minor (and sometimes not so minor) failures to achieve the design goals.

2.5.5 Manufacturing Engineering

The description above relates more perfectly to the one- or two-of-kind system developed by a specialty house. The process is somewhat different for systems that are to be mass produced.

The engineering mass-production effort can be broadly classified into two categories: (1) design engineering and (2) manufacturing engineering. When the design engineers have completed their effort, a small number of prototype models are built and tested to prove the soundness of the design and to allow the design team to refine the drawings and specifications to the point where they can be turned over to the manufacturing division for mass production.

Of course, the design engineers have been working in close liaison with the manufacturing engineers, so that advanced planning for special manufacturing tools or methods could be initiated and so the design engineers could gain information on manufacturing capabilities.

Manufacturing engineers must take additional steps to assure that mass-produced systems will not fail at a high rate. Statistically, the mass-produced units may not equal the performance of the prototype units for a number of reasons. The solution to this problem is to be found in a sound approach to tolerance buildup control, which must be based on the manufacturing machine capabilities.[9]

2.5.6 Documentation

Documentation is essential to the design process; otherwise the investment in engineering time and money may be lost. It has been estimated that more than 50% of an engineer's time is consumed in providing adequate documentation.

Few engineers appreciate the importance of laboratory notebooks.[10] When properly used they can make the engineer's work easier and provide a legal record. The laboratory notebook should provide a continuous record of the design thought process, including blind alleys as well as successes.

The laboratory notebook provides a legal record that may be necessary for patent proof. Many companies require their patent lawyer to inspect the notebooks periodically to assure that they are acceptable. To qualify as a legal record:

1. The notebook should be bound (never use a looseleaf), and the pages should be numbered.
2. All entries should be in ink. Any changes should be crossed out (but not obliterated), and the changes should be initialed and dated.
3. A large cross should be placed through any unused space (to assure that material is not added later).
4. Only one project should be entered on a given page. Some companies provide notebooks that contain a perforated carbon copy duplicate page that the engineer can remove to assemble notebook information pertaining to a given project.
5. Every page of the notebook should be dated, witnessed, and signed. The person witnessing what is on the page should understand it.

Reports are a basic product of an engineering research laboratory. The laboratory notebook provides the resource material for the report.

Production and manufacturing also require engineering reports, which may include such things as maintenance and operation manuals. Fundamental to any manufacturing process are detailed specifications and drawings that describe every part and manufacturing process required to produce and test the system.

2.6 RADIOMETRIC PERFORMANCE EQUATION

The objective of this section is to develop a generalized radiometric performance equation as a unifying theme, specifically for the feasibility study and in general for the entire book. As indicated in Sec. 2.5.1, the overriding radiometric performance parameter required for systems designed for information transfer is the signal-to-noise ratio (SNR).

The SNR equation involves the source, intervening medium, optical subsystem, detector, signal-conditioning optics, and output and display. The complete equation is unmanageably complex until simplifying assumptions are introduced to make it workable. It is a function of the spatial, spectral, temporal, polarization, and coherence of the target, background, and sensor. The performance equation can take on many forms depending upon the simplifying assumptions that must be made. It can be written to show the dependence upon, and sensitivity to, various system parameters.

The radiometric performance equation is the mathematical expression that quantitatively characterizes the SNR, taking into account all of the pertinent factors contributing to the result. Fortunately, the effect of some of the parameters is often negligible; the effects of others are small and easily evaluated. The basic problem is to identify and correctly assess the effects of all significant factors, making simplifications wherever possible in order to obtain a useful expression. Of particular interest is the part of the equation that deals with parameters under the control of the designer. Sensitivity analysis, which has been defined[11] as a calculation of the ratio of target signal to total noise, is facilitated by appropriately simplified performance equations.

In its simplest form, the radiometric performance equation is given by the ratio of the effective flux ϕ_{eff} to the noise equivalent flux (NEF):

$$\text{SNR} = \frac{\phi_{eff}}{\text{NEF}} \tag{2.1}$$

where ϕ_{eff} is that part of the flux incident upon the sensor entrance aperture that is effective in evoking a response in the sensor, and NEF is that level of flux incident upon the sensor entrance aperture that produces an average change in the output signal equal to the root-mean-square (rms) noise.

Generally, a signal-to-noise ratio of 1 is not adequate for purposes of signal conditioning, since the peak-to-peak noise is approximately 3 to 5 times its rms value. Consequently, a minimum SNR of 10 to 100 is often required; however, in spite of this, the use of the rms value is mathematically convenient.

Equation (2.1) can be expressed in terms of energy rate (watts) or photon rate; it can also be written in terms of the appropriate entities used to characterize sources such as sterance [radiance],* which is appropriate for extended-area sources, and pointance [exitance], which is appropriate for point sources. Equation (2.1) can also be used for other radiance entities such as the rayleigh or areance [irradiance].

---

*See Sec. 3.3 for explanation of brackets.

Some of the fundamentals of source characterization, transfer of flux, and the selectivity of optical systems must be developed before even a simplified version of the radiometric performance equation can be written. Chapters 3 and 4 provide the geometrical basis for calculating the flux transferred from the source to the sensor entrance aperture. Chapter 5 develops the fundamentals of spatial and spectral selectivity; Chap. 6 provides detector figures of merit, including detector noise; Chap. 7 gives parameters to characterize the temporal response function (information bandwidth and noise bandwidth). At that point it is possible to write the radiometric performance equation for a specific measurement configuration, which is the subject matter of Chap. 8. Finally, Chap. 9 gives a specific example of a feasibility study.

EXERCISES

1. What is the overriding sensitivity requirement for all electro-optical systems designed for information processing?
2. What is an optimum design in terms of fundamental effects?
3. What level of specifications is established by the feasibility study?
4. How do technical specifications relate to system performance goals?
5. How are complex systems made simple for design purposes?
6. How is the quality of optical measurements related to "spectral purity" and "spatial purity" in connection with electro-optical selectivity?
7. How are out-of-band and off-axis response characteristics of an electro-optical sensor related to sensor selectivity?
8. Classify the functional blocks in Fig. 2.2 in accordance with the generalized electro-optical system illustrated in Fig. 1.1. What systems are missing?
9. Why is the selection of a detector of electro-optical radiant energy so central to sensor system design?

REFERENCES

1. *Webster's Ninth New Collegiate Dictionary*, Merriam-Webster, Springfield, MA, 1983.
2. J. C. Hancock, *Introduction to Electrical Design*, Holt, Rinehart, and Winston, New York, 1972, p. 3.
3. Examples not already cited are: J. R. M. Alger and C. V. Hayes, *Creative Synthesis in Design*, Prentice-Hall, Englewood Cliffs, NJ, 1967; E. V. Krick, *An Introduction to Engineering and Engineering Design*, Wiley, New York, 1965.
4. L. A. Wainstein and V. D. Zubakov, *Extraction of Signals from Noise*, Prentice-Hall, Englewood Cliffs, NJ, 1962, p. 137.
5. O. R. Wade, *Tolerance Control in Design and Manufacturing*, Industrial Press, New York, 1967, p. v.
6. T. T. Furman, *Approximate Methods in Engineering Design*, Academic Press, New York, 1981, p. 50.
7. F. E. Nicodemus, "Optical Instruments," Part I, in R. Kingslake, Ed., *Applied Optics and Optical Engineering*, Vol. 4, Academic Press, New York, 1967, p. 288.

8. B. S. Blanchard and W. J. Fabrycky, *Systems Engineering and Analysis*, Prentice-Hall, Englewood Cliffs, NJ, 1981, p. 240.

9. O. R. Wade, *Tolerance Control in Design and Manufacturing*, Industrial Press, New York, 1967, p. 173.

10. R. V. Hughson, "The Right Way to Keep Laboratory Notebooks," *New Engineer*, **4**, May 1975, pp. 26–31.

11. W. L. Wolf, Ed., *Handbook of Military Infrared Technology*, Office of Naval Research, Washington, DC, 1965, p. 731.

chapter *3*

Radiant Sources

3.1 INTRODUCTION

In the development of subsystem figures of merit for use in the radiometric performance equation, it is necessary to find the effective flux transferred from the source to the receiver. The methodology of Chaps. 3 through 7 is to trace the signal from the source to the output and display subsystem. That signal consists of an optical beam of flux, encoded with information of value to the system user, that results in flux on the detector. From that point the signal is the electric output of the detector that must be amplified, decoded (dc restored), and filtered to produce a usable output signal.

The objective of this chapter is to model the source of optical flux in terms of geometrical radiant entities that serve as source figures of merit to be used in the radiometric performance equation. The figures of merit permit optimization of the system in terms of such factors as sterance [radiance], wavelength, temperature, and size of the source.

These geometrical entities provide useful and practical means to predict radiant energy transfer.[1] The characterization of radiant sources in terms of Planck's equation is given in Chap. 10.

The transfer of radiant flux through space is evaluated using geometrical concepts of area, projected area, angle, solid angle, projected solid angle, and beams (or rays) of optical energy. The geometrical concepts are introduced (Sec. 3.2), after which the radiant entities are defined (Sec. 3.3).

Table 3.1 GEOMETRICAL ENTITIES

Terms	Relations[a]	Units
Projected area	$A_p = \int_A \cos \theta \, dA$	cm^2
Solid angle	$\omega = \int_0^{2\pi} d\phi \int_0^{\Theta} \sin \theta \, d\theta = 2\pi(1 - \cos \Theta)$	sr
Projected solid angle	$\Omega = \int_0^{2\pi} d\phi \int_0^{\Theta} \sin \theta \cos \theta \, d\theta = \pi \sin^2 \Theta$	sr
Throughput	$\Upsilon_1 = \int_{A_1} \int_{\omega_1} dA_1 \cos \theta \, d\omega_1$	cm^2 sr

[a]It is assumed that the spherical Z axis is coincident with the axis of the right circular cone.

3.2 GEOMETRICAL ENTITIES

The geometrical entities of interest in this section are listed in Table 3.1.

3.2.1 Projected Area

The *area* of a rectilinear projection of a surface (not necessarily a plane surface) onto a plane perpendicular to the unit vector $\hat{\mathbf{k}}$ is the *projected area* A_p and is given by

$$A_p = \int_A \cos \theta \, dA \qquad [cm^2] \qquad (3.1)$$

where θ is the polar angle between the unit vector $\hat{\mathbf{k}}$ and the normal to the element dA of the surface. This can be visualized in terms of an element of the surface of a hemisphere projected onto the base of the hemisphere as illustrated in Fig. 3.1.

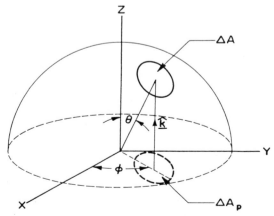

Figure 3.1 Illustration of projected area.

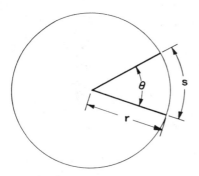

Figure 3.2   Geometry for radian measure of angle.

3.2.2   Degrees and Radians

Degrees and *radians* are measures of an angle in a plane surface. As illustrated in Fig. 3.2, the *radian* (rad) is defined in reference to a circle as the ratio of the segment s to the radius r. The radian measure of a full circle is given by the ratio of the circumference $2\pi r$ divided by the radius r, which yields 2π. In other words, 2π rad $= 360°$.

3.2.3   Solid Angle and Projected Solid angle

The *steradian* (sr) is a measure of a *solid angle* (an angle in a volume) as illustrated in Fig. 3.3. The steradian is defined in reference to a sphere (or hemisphere) as the ratio of the area A intercepted on the surface of the sphere to the square of the radius r. Thus the differential solid angle in spherical coordinates is given by

$$d\omega = \frac{dA}{r^2} = \sin\theta \, d\theta \, d\phi \qquad [\text{sr}] \qquad (3.2)$$

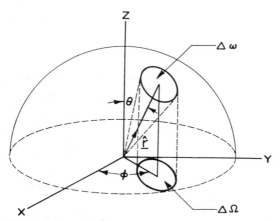

Figure 3.3   Illustration of solid angle and projected solid angle.

A more general definition is given by

$$dw = \frac{\hat{\mathbf{r}} \cdot dA}{r^2} \quad [\text{sr}] \quad (3.3)$$

The dot product gives the projected area normal to the unit vector $\hat{\mathbf{r}}$ (see Fig. 3.3). The dot product is everywhere satisfied in Eq. (3.2) provided the vertex of the solid-angle cone is at the center of the sphere.

The *solid angle* for a right circular cone oriented with its center on the Z axis is given by

$$\omega = \int_0^{2\pi} d\phi \int_0^{\Theta} \sin \theta \, d\theta = 2\pi (1 - \cos \Theta) \quad [\text{sr}] \quad (3.4)$$

where Θ is the half-angle of the solid angle. This is illustrated in Fig. 3.4

The element of projected solid angle is given by

$$d\Omega = \cos \theta \, d\omega = \cos \theta \sin \theta \, d\theta \, d\phi \quad [\text{sr}] \quad (3.5)$$

where θ is the angle between the cone axis and the zenith axis of the unit hemisphere (see Fig. 3.3). It may be visualized as the projection of the area (solid angle $d\omega$) of the unit hemisphere onto the base of the hemisphere. The *projected solid angle* Ω for a right circular cone oriented with its center on the Z axis is given by

$$\Omega = \int_0^{2\pi} d\phi \int_0^{\Theta} \sin \theta \cos \theta \, d\theta = \pi \sin^2 \Theta \quad [\text{sr}] \quad (3.6)$$

where Θ is the half-angle edge of the solid angle (see Fig. 3.4). The value of Ω is equal to ω to within 1% for half-angles less than $\Theta = 10°$.

For a full hemisphere ($\Theta = 90°$), Eq. (3.4) yields 2π sr, which is one-half the area of a unit sphere, while Eq. (3.6) yields π, which is the area of the base of a unit hemisphere.

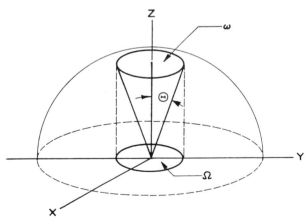

Figure 3.4 Illustration of solid angle as a right circular cone with its center aligned along the Z axis.

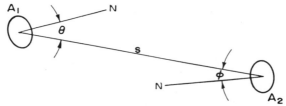

Figure 3.5   Geometry of a ray.

3.2.4   The Optical Beam and Throughput

A beam or ray of optical energy is defined in terms of two finite areas[2] as illustrated in Fig. 3.5. Such a beam consists of the geometrical region that is bounded at the two nonzero cross-sectional areas A_1 and A_2 that are separated by a distance s. All the flux in the beam flows through A_1 and A_2.

The optical beam can be quantified in terms of the basic throughput, which is defined by

$$\Upsilon = n^2 A \Omega \qquad [\text{cm}^2 \text{ sr}] \qquad (3.7)$$

where n is the index of refraction, A is the area, and Ω is the projected solid angle.

The throughput[3] for a homogeneous medium where $n = 1$ for the beam of Fig. 3.5 is given for A_1 as

$$\Upsilon_1 = A_1 \cos \theta \omega_1 = A_1 \cos \theta \, \frac{A_2 \cos \phi}{s^2} \qquad [\text{cm}^2 \text{ sr}] \qquad (3.8)$$

and is given for A_2 by

$$\Upsilon_2 = A_2 \cos \phi \omega_2 = A_2 \cos \phi \, \frac{A_1 \cos \theta}{s^2} \qquad [\text{cm}^2 \text{ sr}] \qquad (3.9)$$

Examination of Eqs. (3.8) and (3.9) shows that $\Upsilon_1 = \Upsilon_2$, which illustrates the *invariance of throughput*. Because the throughput is invariant, it can be applied to a source, an entrance aperture, a stop, a detector, or to any abstract point along a beam. When we say the throughput is invariant we mean that it has the same numerical value at any point along the beam.[4]

The above results are true only in the limit as the distance s becomes large compared with the linear dimensions of A_1 and A_2. The exact value of the throughput at A_1 is given by

$$\Upsilon_1 = \int_{A_1} \int_{\omega_1} dA_1 \cos \theta \, d\omega_1 \qquad [\text{cm}^2 \text{ sr}] \qquad (3.10)$$

The integral can be evaluated provided ω_1 is independent of A_1, which occurs when s is large compared to the linear dimension of A_1. Then the throughput is given by Eq. (3.8). There is a rule of thumb that states that Eqs. (3.8) and (3.10) agree to within about 1% when s is 20 times the maximum linear dimension of A_1.

3.2.5 Configuration Factors

There are means to obtain an exact evaluation of Eq. (3.10) when the approximations given above are not adequate. The factors used to obtain the solution are known by a variety of names: angle factor, angle ratio, geometrical factor, interchange factor, shape factor, form factor, exchange coefficient, and configuration factor.[5] Configuration factors are widely used in engineering literature, and there are extensive tables used in evaluating them.[6,7]

The configuration factor F_1 can be defined in reference to Fig. 3.6 as the ratio of the throughput in the beam at A_1 to the hemispheric throughput of A_1 using Eq. (3.6) for $\Theta = 90°$ and Eq. (3.10):

$$F_1 = \frac{\Upsilon_1}{\Upsilon_{\text{hem}}} = \frac{\int_{A_1}\int_{\omega_1} dA_1 \cos\theta \, d\omega_1}{\pi A_1} \tag{3.11}$$

The throughput Υ_1 is therefore given by

$$\Upsilon_1 = F_1 \pi A_1 \tag{3.12}$$

A particularly useful case for electro-optical system design is illustrated in Fig. 3.6 for two concentric and parallel circularly symmetrical cross-sectional areas. They exhibit a relatively large radius relative to their separation. This orientation occurs in optical systems where the areas might represent lenses or stops.

The throughput $A_1\Omega_1$ is given by

$$\Upsilon_1 = F_1 \pi A_1 \tag{3.13}$$

where

$$F_1 = \tfrac{1}{2}\left(z - \sqrt{z^2 - 4x^2 y^2}\right) \tag{3.14}$$

and where

$$x = \frac{r_2}{s}, \qquad y = \frac{s}{r_1}, \qquad z = 1 + (1 + x^2)y^2 \tag{3.15}$$

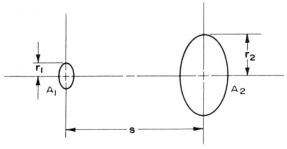

Figure 3.6 Geometry for parallel circularly symmetrical area throughput calculation using configuration factors.

Example 1:   Find the exact and approximate throughput Υ_1 for a parallel circularly symmetrical configuration like that of Fig. 3.6. Compare the results.

Given: $r_1 = 1$, $r_2 = 2$, and $s = 20$ cm.

Basic equations:

$$\Upsilon_1 = A_1 \cos\theta \frac{A_2 \cos\phi}{s^2} \tag{3.8}$$

$$\Upsilon_1 = F_1 \pi A_1 \tag{3.13}$$

$$F_1 = \tfrac{1}{2}\left(z - \sqrt{z^2 - 4x^2 y^2}\right) \tag{3.14}$$

$$x = \frac{r_2}{s}; \quad y = \frac{s}{r_1}; \quad z = 1 + (1 + x^2)y^2 \tag{3.15}$$

Assumptions: None

Solution: The approximate solution is given by Eq. (3.8), where $\cos\theta = \cos\phi = 1$ and

$$\frac{A_1 A_2}{s^2} = \frac{\pi \times 1^2 [\text{cm}^2] \times \pi \times 2^2 [\text{cm}^2]}{20^2 [\text{cm}^2]} = 9.8696 \times 10^{-2} \text{ cm}^2 \text{ sr}$$

The exact solution is given by Eqs. (3.13) through (3.15):

$$x = \frac{2}{20} = 0.1 \qquad y = \frac{20}{1} = 20$$

$$z = 1 + \left[1 + (0.1)^2\right] \times 20^2 = 405$$

$$F_1 = \tfrac{1}{2}\left[405 - \left[405^2 - 4(0.1)^2 \times 20^2\right]^{1/2}\right] = 9.8772 \times 10^{-3} \quad \text{(unitless)}$$

$$\Upsilon = 9.8772 \times 10^{-3} \times \pi \times \pi \times 1^2 = 9.7483 \times 10^{-2} \text{ cm}^2 \text{ sr}$$

The approximate and exact solutions differ by 1.24%, which illustrates the rule of thumb given in Sec. 3.2.4.   ∎

3.3   BASIC RADIANT ENTITIES

One of the major problems in the field of electro-optics is that of nomenclature. Many different systems of nomenclature of terms, symbols, and units exist. This multiplicity of terminology discourages interdisciplinary communication and adds to the confusion.

The standards that probably have achieved the widest acceptance are those developed by the International Commission on Illumination (CIE).[8] They have been adopted by the National Bureau of Standards[9] and other American societies and symposia.[10] Because of their acceptance by these standardizing agencies and because of their wide use, the CIE standards are used in this book. However, they are incomplete.

New terms have been suggested[11] to supplement the CIE terms. These new terms are designed to suggest the geometrical characteristic of the

Table 3.2 BASIC RADIOMETRIC ENTITIES

Terms	Symbol	Units[a]
Flux (general)	Φ	ϕ
Radiant flux	Φ_e	W
Photon flux	Φ_p	q/s
Luminous flux	Φ_v	lm
Spectral flux	Φ_λ	$\phi/\mu m$
Areance [exitance]	M	ϕ/cm^2
Sterance [radiance]	L	$\phi \; cm^{-2} \; sr^{-1}$
Pointance [intensity]	I	ϕ/sr
Areance [irradiance]	E	ϕ/cm^2

[a]ϕ is a general term for unit of flux. $q =$ quantum.

entity. They are

> Sterance: related to solid angle (steradian)
>
> Areance: related to area
>
> Pointance: related to a point

These new terms are used in this book with the CIE term, when defined, immediately following in square brackets.

The international system[9] of units (SI) is used for dimensional analysis. However, some exceptions in common use are noted. Appendix A gives a list of the entities of interest in this book.

The subscripts e, p, and v, used in Table 3.2, refer to *energy*, *photon* (or quanta), and *visible*, respectively. They are not used when it is clear from the context.

The confusion related to terms used can be greatly alleviated through the use of the associated units. Any radiometric term or symbol is most fundamentally defined in terms of units. For example, the units of the expression $\Phi = 1.6 \; W/cm^2$ clearly identify Φ as radiant *areance*, which has the units of watts per square centimeter. In the CIE standard, there are two terms, *exitance* and *irradiance*, associated with these units.

Thus, the units are given following each expression throughout this book wherever it is appropriate. This should result in a more readable text, especially for those not familiar with the terms and/or symbols used.

3.3.1 Flux

Flux, Φ, is defined as any quantity that is propagated or spatially distributed according to the laws of geometry. Examples are radiant energy and power, visible light, infrared radiation, quanta or photons, and spectral radiant power. The term *flux* is a general term to be used when the parameters being considered are independent of the units associated with the flux. In this case the symbol ϕ is used to denote the units for general flux Φ as given in Table 3.2.

Radiant flux, photon flux, and spectral flux are more specific terms that imply units of energy rate, quanta rate, and spectral density as given in Table 3.2.

3.3.2 Spectral Flux

Spectral flux is an entity that can be defined as differential with respect to wavelength or optical frequency. For example, the spectral flux

$$\Phi(\lambda) = \frac{d\Phi}{d\lambda} \qquad [\phi/\mu m] \tag{3.16}$$

is the ratio of the radiant flux to the wavelength interval $\Delta\lambda$ as $\Delta\lambda$ is reduced to a particular wavelength and has the units $\phi/\mu m$. *Note:* It is common to express wavelength in micrometer (1 $\mu m = 10^{-6}$ m) rather than in meters. The value of $\Phi(\lambda)$ will, in general, vary with wavelength. The total flux between the wavelengths λ_1 and λ_2 is given by

$$\Phi = \int_{\lambda_1}^{\lambda_2} \Phi(\lambda)\, d\lambda \tag{3.17}$$

3.3.3 Sterance [Radiance]

One might expect that the radiation per unit solid angle (in a specific direction) from a flat zero-thickness surface element varies with the projected area $A_p = A\cos\theta$ in that direction. Both the projected area and flux per unit solid angle vary with the cosine of the polar angle θ:

$$\Phi = \Phi_0\cos\theta \tag{3.18}$$

where Φ_0 is the flux per unit solid angle and θ is the angle between the surface normal and the ray direction. This is known as *Lambert's cosine law* (see Fig. 3.7) and explains why a spherical body with relatively uniform surface conditions, such as the sun, appears as a flat disk. The projected

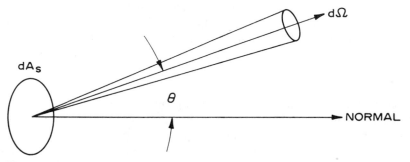

Figure 3.7 Pointance [intensity] in a direction with respect to the normal.

area near the limb represents more real area, but that area is radiating less in the oblique direction so the effects cancel and the disk looks uniformly bright. Lambert's law holds only for the diffuse component of the flux. The specular (reflective) component obeys Snell's law of reflection.

When the geometrical distribution of radiation from a surface obeys Lambert's cosine law, the source is said to be Lambertian, or perfectly diffuse.

Sterance [radiance] is the most general term to describe source flux because it includes both positional and directional characterization. The sterance [radiance] is defined as

$$L = \frac{d^2\Phi}{dA \cos\theta \, d\omega} \qquad [\phi \ cm^{-2} \ sr^{-1}] \qquad (3.19)$$

where $d\omega$ is the solid angle into which the flux is being radiated and $\cos\theta \, dA$ is the projected source area. The sterance [radiance] is the ratio of the flux to the product of the area and the solid angle as the area is reduced to a point (positional) and as the solid angle is reduced to a specific direction (directional). (*Note:* The definition of sterance is based on radiation from only one side of the source area.)

The visual equivalent of sterance may be termed "brightness." The definition of sterance in terms of projected area makes the sterance or brightness independent of direction. This is true only if the source is perfectly diffuse or Lambertian but is common in nature and explains why a photographic exposure is independent of distance or direction except when specular reflectance occurs as on water or snow.

The total flux is given by

$$\Phi_T = \int_A \int_\omega L \, dA \cos\theta \, d\omega \qquad [\phi] \qquad (3.20)$$

The invariance theorem[4] states that the sterance [radiance] is constant within a homogeneous environment. Thus the flux has the same numerical value at the sensor entrance aperture as it does at the source. This leads to the conclusion that sterance [radiance] is the most appropriate entity to characterize the flux for an extended-area source. The characterization "extended-area source" applies to any source that completely fills the sensor field of view.

Example 2: Find the sterance [radiance] of a nuclear burst fireball.

Given: Radius $r = 2.5$ km, total power $\Phi_e(T) = 10^9$ W.

Basic equation:

$$L = \frac{d^2\Phi}{dA \cos\theta \, d\omega} \qquad (3.19)$$

Assumption: The *average* sterance [radiance] is equal to the sterance [radiance] for the case where the radiation is uniform and isotropic (independent of direction).

Solution: The average sterance [radiance] is given as the ratio of the total flux to the product of the solid angle 4π sr (for a sphere) and the projected area (the disk).

$$L(\text{ave}) = \frac{10^9[\text{W}]}{4\pi[\text{sr}] \times \pi(2.5 \times 10^3[\text{m}])^2 \times 1 \times 10^4[\text{cm}^2/\text{m}^2]}$$

$$= 4.05 \times 10^{-4} \text{ W cm}^{-2} \text{ sr}^{-1}$$

Equation (3.20) can be written as

$$\Phi = L \int_A \int_\omega dA \cos\theta \, d\omega = L\Upsilon \qquad [\phi] \qquad (3.21)$$

where Υ is the throughput given by Eq. (3.10) provided L and ω are independent of position. For many practical cases the throughput can be given by Eq. (3.8) in accordance with the rule of thumb given above, so that the flux in a beam from a source can be written as

$$\Phi = L A_1 \cos\theta \frac{A_2 \cos\phi}{s^2} \qquad (3.22)$$

in reference to Fig. 3.5.   ∎

Example 3:   Find the flux in watts radiated into a solid angle of 1×10^{-3} sr at an angle of $\theta = 30°$ to the normal of a flat-plate radiator of area = 1 cm².

Given: The flat plate is radiating $L = 1800$ W cm⁻² sr⁻¹.

Basic equation:

$$\Phi = L \int_A \int_\omega dA \cos\theta \, d\omega = L\Upsilon \qquad (3.21)$$

Assumptions: The surface properties of the flat plate are assumed to be Lambertian or perfectly diffuse, so the radiation in any direction is proportional to the cosine of the angle with the normal to the surface. These assumptions are inherent with the definition of sterance [radiance], Eq. (3.19). It is also assumed that the flat plate is a uniform radiator; that is, it has the same value of sterance [radiance] at each point.

Solution: The flux in the beam is given as the product of the through-put and the sterance [radiance] in the beam calculated at the surface of the flat plate.

The throughput can be calculated as the product of the projected area (in the direction θ) and the solid angle:

$$\Upsilon = 1[\text{cm}^2] \times \cos 30° \times 1 \times 10^{-3}[\text{sr}] = 8.66 \times 10^{-4} \text{ cm}^2 \text{ sr}$$

The flux in the beam is

$$\Phi = 8.66 \times 10^{-4}[\text{cm}^2 \text{ sr}] \times 1800[\text{W cm}^{-2} \text{ sr}^{-1}] = 1.6 \text{ W} \qquad \blacksquare$$

3.3.4 Areance [Exitance]

The areance [exitance] characterizes the source in terms of position only and is defined as

$$M = \frac{d\Phi}{dA} \qquad [\phi/\text{cm}^2] \qquad (3.23)$$

The areance [exitance] is the ratio of the flux to the source area from which it is radiating as the area is reduced to a point (positional).

The total flux radiated into a hemisphere (one side of surface only) is given by

$$\Phi = \int M \, dA \qquad [\text{W}] \qquad (3.24)$$

3.3.5 Relationship between Sterance and Areance

The relationship between sterance [radiance] and areance [exitance] of a source is obtained as follows: Consider an incremental radiating surface ΔA, small compared to a hemisphere of radius r, that has a sterance L as illustrated in Fig. 3.8. The flux upon the incremental collecting area dA_c is

$$d\Phi = L \Delta A_s \frac{\cos \theta \, dA_c}{r^2} \qquad [\phi] \qquad (3.25)$$

where the incremental projected solid angle is

$$d\Omega = \frac{\cos \theta \, dA_c}{r^2} \qquad (3.26)$$

The total flux radiated by ΔA_s is obtained by integration over the entire hemisphere:

$$\Phi_T = L \Delta A_s \int d\Omega = L \Delta A_s \pi \qquad [\phi] \qquad (3.27)$$

The total flux is also given by

$$\Phi = M \Delta A_s \qquad (3.28)$$

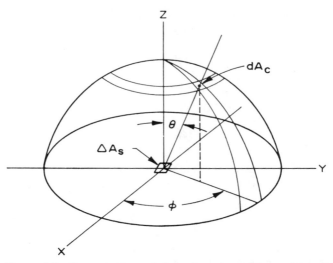

Figure 3.8   Geometry for radiation of an elemental area into a hemisphere.

so that

$$M = L\pi \quad [\phi/cm^2] \tag{3.29}$$

3.3.6 Pointance [Intensity]

Pointance [intensity] characterizes the source in terms of the direction only and is defined as

$$I = \frac{d\Phi}{d\omega} \quad [\phi/sr] \tag{3.30}$$

The pointance [intensity] is the ratio of the flux to the solid angle as the solid angle is reduced in value about a specific direction (directional) and can be written as

$$I = I_0 \cos\theta \tag{3.31}$$

where I_0 is the pointance [intensity] in a direction normal to the surface.

The total flux is given by

$$\Phi_T = \int_\omega I\,d\omega = I_0\int \cos\theta\,d\omega \tag{3.32}$$

for a flat surface. The pointance may be appropriately used to characterize a point source, since its area may be unknown. Equation (3.31) does not apply for an isotropic point source. In this case the total flux is given by

$$\Phi_T = \int I\,d\omega = 4\pi I \tag{3.33}$$

since it radiates into a sphere.

3.3.7 Areance [Irradiance]

Areance [irradiance] is a measure of the total incident flux per unit area upon a receiver and is defined as

$$E = \frac{d\Phi}{dA} \qquad [\phi/\text{cm}^2] \tag{3.34}$$

The areance [irradiance] is the ratio of the incident power to the area upon which it is incident as the area is reduced to a specific position (positional).

The total flux is given by

$$\Phi_T = \int_A E \, dA \qquad [\text{W}] \tag{3.35}$$

The areance [irradiance] at the sensor aperture is an appropriate entity to characterize a point source when the distance to the source is unknown.

The flux density that is incident upon a sensor aperture A_c from an isotropic point source is given by

$$E = \frac{\Phi_T}{A_c} = \frac{1}{A_c} \int I \, d\omega = \frac{I}{s^2} \tag{3.36}$$

since $\omega = A_c/s^2$ and s is the distance between the source and the sensor.

Example 4: Find the areance [irradiance] of the fireball given in Example 2 at the range of 50 km.

Basic equation:

$$I = d\Phi/d\omega \tag{3.30}$$
$$E = I/s^2 \tag{3.36}$$

Assumptions: The same assumptions hold as given in Example 2: the source is uniform and isotropic. In addition, it is assumed that the intervening medium has no effect upon the radiation incident at a distance.

Solution: The pointance [intensity] is equal to the average pointance, which is given as the ratio of the total flux to the solid angle (of a sphere) 4π sr.

$$I(\text{ave}) = 10^9 [\text{W}]/4\pi [\text{sr}] = 7.96 \times 10^7 \text{ W/sr}$$

The areance [irradiance] is given by the ratio of the pointance to the distance squared:

$$E = \frac{7.96 \times 10^7 [\text{W/sr}]}{\{50 \times 10^5 [\text{cm}]\}^2} = 3.18 \times 10^{-6} \text{ W/cm}^2$$

Note: The unit steradian (sr) is really length squared divided by length squared, and care must be exercised in using it. In this case the quantity

$1/s^2$ can have the units of steradians if it is viewed as the solid angle subtended by unit area at the distance s.   ■

3.3.8 Radiant Field Quantities

The terms sterance, areance, and pointance were defined as having reference to a source or a receiver. However, these concepts can also be applied within a radiation field away from sources or receivers. If a barrier containing an aperture is placed in a radiant field, it has the properties of a source for the flux leaving the aperture and the properties of a receiver for the flux incident upon it. The radiation field could be defined by reducing the aperture to a point; then there would be a meaningful measure of the sterance, areance, and pointance for that point or of the field. Thus, there is no fundamental reason for distinguishing between incoming or outgoing flux for any entity. The utility of considering these entities as field quantities will become evident in later sections.

3.4 BASIC PHOTOMETRIC ENTITIES

Photometry is the study of the transfer of radiant energy in the form of light. Rather than measuring light in terms of energy rate or photon rate, the visual effect is taken into account. The visual sensitivity of the *standard observer* is described in terms of the response of the standard human eye. The field of photometry is historically concerned with lighting conditions or *illumination* in the design of buildings for human occupation.

There exists a direct correspondence among photometric and radiometric units, as shown in Table 3.2. For example, luminous sterance has the units of lumens per square centimeter per steradian (lm cm^{-2} sr^{-1}). The CIE term for luminous sterance is *luminance*, and *photometric brightness* is an alternative term.

The luminous areance, with units of lm/cm^2, has the CIE term of *illuminance*, and *illumination* is an alternative. Luminous areance is used with units of footcandle (lumen per square foot), or lux (lumen per square meter), or stilb (lumen per square centimeter), or nit (candela per square meter), or foot lambert ($1/\pi$ candela per square foot), or apostilb ($1/\pi$ candela per square meter).

The luminous pointance, with units of lumens per steradian (lm/sr), has the CIE term of *luminous intensity* with *candlepower* as an alternative. Luminous pointance is used with the unit *candela* (formerly candle), which is equivalent to 1 lumen per unit solid angle.

3.5 BASIC PHOTON FLUX ENTITIES

No terms are given in the CIE nomenclature for the entities that deal with photon flux. However, the same general scheme given in Table 3.2 applies for the geometric concepts of sterance, areance, and pointance.

The photon sterance has units of $q\,s^{-1}\,cm^{-2}\,sr^{-1}$, the photon areance has units of $q\,s^{-1}\,cm^{-2}$, and the photon pointance has units of $q\,s^{-1}\,sr^{-1}$, where q stands for quantum or quanta.

EXERCISES

1. Prove that $M = L\pi$ for a hemisphere by carrying out the indicated integration. (Refer to Sec. 3.3.5, Fig. 3.7, and Table 3.1.)
2. The measured areance [irradiance] E for a distant star is 3.3×10^{-11} W/cm^2. The star is located 1.3 light-years from earth. Find the apparent pointance [intensity] of the star.
3. A flat-plate radiator is emitting 3.8 W/cm^2. Find its sterance [radiance] in units of W cm^{-2} sr^{-1}.
4. Given a 4-cm-diameter circular flat-plate radiator for which the total power radiated into a hemisphere is $\Phi = 1.5 \times 10^{-2}$ W. Describe the source in terms of (a) sterance [radiance], (b) areance [exitance], and (c) pointance [intensity] normal to the surface. List the assumptions required for a solution.
5. Given a spherical isotropic source of diameter 4 cm for which the total power radiated into a sphere is $\Phi = 6.3 \times 10^{-4}$ W. Describe the source in terms of (a) sterance [radiance], (b) areance [exitance], and (c) pointance [intensity]. List the assumptions required for a solution.
6. Find the solid angle and the projected solid angle for a circularly symmetrical cone coincident with the spherical Z axis for $\theta = 1, 5, 10, 20$, and 90°. (See Table 3.1 and Fig. 3.4.)
7. Calculate the hemispheric throughput for a 1-cm diameter circular aperture.
8. Calculate the exact and approximate throughput for a circularly symmetrical configuration like that of Fig. 3.6, where $r_1 = 1$, $r_2 = 3$, and $s = 15$ cm. Calculate the percent error for the approximate solution.
9. A flat plate radiates as a perfectly diffuse (Lambertian) radiator $M = 1.6$ W/cm^2 (Fig. E3.9). What is its sterance L, in W cm^{-2} sr^{-1}, in the direction θ?
10. The sterance L_s of a 1-cm^2 flat plate is 500 W cm^{-2} sr^{-1} (Fig. E3.10). Find the flux radiated into $\omega = 10^{-3}$ sr at $\theta = 30°$.

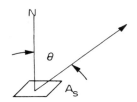

Figure E3.9

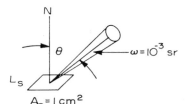

Figure E3.10

11. An isotropic point source (spherical) radiates $\Phi = 1.6 \times 10^3$ W. Find the average pointance I in W/sr.

12. A perfectly diffuse flat-plate source radiates a total of $\Phi_T = 21$ W into a hemisphere. Its area is 5 cm². Find the average sterance in W cm^{-2} sr^{-1}.

REFERENCES

1. F. E. Nicodemus, Ed., "Self-Study Manual on Optical Radiation Measurements. Part 1—Concepts," *Natl. Bur. Stand. (U.S.), Tech. Note*, No. 910-1, 1976, p. 82.
2. Ibid., p. 10.
3. W. H. Steel, "Luminosity, Throughput, or Entendue? Further Comments." *Appl. Opt.*, **14**, 252 (1974).
4. F. E. Nicodemus, "Radiance," *Am. J. Phys.*, **31**, 368 (1963).
5. F. E. Nicodemus, Ed., "Self-Study Manual on Optical Radiation Measurements: Part 1—Concepts," *Natl. Bur. Stand. (U.S.), Tech. Note* No. 910-2, 1976, pp. 93–100.
6. R. Siegel and J. R. Howell, *Thermal Radiation Heat Transfer*, McGraw-Hill, N.Y., 1972.
7. E. M. Sparrow and R. D. Cess, *Radiation Heat Transfer*, Brooks/Cole, Belmont, CA, 1966.
8. *International Lighting Vocabulary*, 3rd ed., Publ. CIE No. 17 (E-1.1), common to the CIE and IEC, International Electrotechnical Commission (IEC), International Commission on Illumination (CIE), Bur. CIE, Paris, 1970.
9. C. H. Page and P. Vigoureux, Eds., "The International System of Units (SI)," *Natl. Bur. Stand. (U.S.), Spec. Publ.* No. 330 (1974).
10. "Optical Society of America Nomenclature Committee Report," *J. Opt. Soc. Am.*, **57**, 854 (1967).
11. I. J. Spiro, "Radiometry and Photometry," *Opt. Eng.*, **13** (1974); **14** (1975); **15** (1976) is a column in which proposals are aired in each issue; see esp. vol. **13**, 1974, pp. G183-G187, and vol. **15**, 1976, p. SR-7.

chapter *4*

Transfer of Radiant Flux

4.1 INTRODUCTION

A knowledge of the effective flux transferred from a source through the intervening medium to a receiver or collector is required to enable one to write the radiometric performance equation.

In the previous chapter, sources were characterized in terms of geometric entities to facilitate calculation of flux transferred from a source. In this chapter the effect of the intervening medium is introduced. This chapter emphasizes path losses through the medium and the *geometry* of radiation transfer, both of which are described in terms of general parameters or figures of merit.

The transfer of radiant energy from a source is described in terms of the radiant entities given in the previous chapter and the effects of the intervening medium. In each case the flux is obtained by integration over the appropriate variables.

4.2 THE INTERVENING MEDIUM

The calculations of the radiometric characteristics of an observed source in an attenuating medium from measurements made at a distance always involve assumptions about the nature of the absorptions, emissions, and scattering of the radiation within the intervening medium.[1]

The effect of the intervening medium is to introduce errors in the calculated values unless corrections are made for them. In the absence of such corrections, the pointance [intensity] of a star, calculated from measurements made on the earth's surface with no correction for the atmospheric attenuation, would be reported as the "apparent pointance [intensity]." On the other hand, the calculations of the total overhead sterance [radiance] of the sky are defined to include the effects of emission, absorption, and scattering and therefore extend from the sensor outward. In this case, the word "apparent" is not appropriate.

The incident areance [irradiance] at the aperture of a sensor is by definition the flux density falling on the sensor aperture; therefore, the word "apparent" is never an appropriate modifier for incident areance or irradiance.

The optical losses that occur at the boundary between two media—for example, mirrors, lenses, and windows—can be expressed as follows:

The relationship

$$\Phi_i = \Phi_\alpha + \Phi_\rho + \Phi_\tau \tag{4.1}$$

is a statement of the conservation of flux (or energy). Equation (4.1) states that the incident flux Φ_i is equal to the sum of the absorbed flux Φ_α, the reflected flux Φ_ρ, and the transmitted flux Φ_τ. Dividing both sides of Eq. (4.1) by Φ_i yields

$$1 = \alpha + \rho + \tau \tag{4.2}$$

The terms α, ρ, and τ are the absorptance, reflectance, and transmittance, respectively.

Terms that end in *-ivity* refer to the ideal property of a material having planar surfaces between two media and no oxides or coatings on the surface or the property of a scattering or absorbing media along a path. Terms, like those above, ending in *-ance* refer to the property of an actual sample or path. Thus, we may speak of the transmittance or absorptance of a filter or the reflectance of a mirror as the unitless ratio of flux transmitted, reflected, or absorbed to the incident flux for a specific path.

Means are available to calculate the spectral transmittance of an atmospheric path to an accuracy of 5 to 10%. Software packages known as LOWTRAN and Aggregate are considered generally useful for a wide variety of problems involving the spectral transmittance from 1 to 30 μm.[2] Both LOWTRAN and Aggregate make use of mathematical models based upon molecular absorption.

The LOWTRAN method, developed by the U.S. Air Force Geophysics Laboratory (AFGL), is more empirically derived and simpler to use than the Aggregate method, although it is somewhat less accurate. LOWTRAN has evolved through a number of versions.

The Aggregate method uses a number of models. Each one is used over that range for which it yields the highest accuracy. To calculate the atmospheric spectral transmittance, the appropriate atmospheric parameters are entered into the formulas. Once the spectral transmittance is known, the spectral sterance [radiance] can be calculated. All calculations are made assuming a flat earth and standardized, horizontally uniform atmospheric conditions. Calculations for a slant path use the so-called equivalent absorber path method, which employs the same formula as for uniform conditions. For slant paths where the angle with respect to the zenith is less than about 80°, the calculation is made for the vertical path and corrected by dividing by the cosine of the angle. Use of the flat-earth assumption introduces errors for horizontal or near-horizontal paths unless the path is relatively short.

The calculations provided by LOWTRAN and/or Aggregate are beyond the scope of this text.

4.3   TRANSFER OF RADIANT ENERGY FROM A SOURCE

The source may be idealized as either a point source or a uniform extended-area source for a first-order analysis. The path transmittance must be included to take into account the path losses. The entity of sterance [radiance] is most appropriate for the description of radiant flux from an extended-area source,[3] which is considered next.

4.3.1   Extended-Area Source

The calculation of the geometrical transfer of flux from an extended-area source to a sensor entrance aperture can be based upon the definition of sterance [radiance] L. This is illustrated with reference to Fig. 4.1 as follows: Dimensional analysis indicates that the flux Φ incident upon the collector of a sensor is obtained by taking the product of the source sterance L, the source projected area A_s, and the solid angle ω_s into which the flux is radiating. This solid angle is defined as the solid angle subtended by the sensor entrance aperture at the source. The differential flux is obtained starting with the definition of spectral sterance [radiance]

$$d^3\Phi = L(\lambda)\, dA_s \cos\theta\, d\omega_s\, d\lambda \qquad (4.3)$$

The total flux is obtained by integrating over the source area, the solid angle (beam) into which the source is radiating, and the wavelength:

$$\Phi_T = \tau_p \int_\lambda \int_{A_s} \int_{\omega_s} L(\lambda)\, dA_s \cos\theta\, d\omega_s\, d\lambda \qquad (4.4)$$

Equation (4.4) can be written as

$$\Phi = \tau_p L(\lambda) \int_\lambda d\lambda \int_{\omega_s} \cos\theta\, d\omega_s \int_{A_s} dA_s = LA_s\Omega_s\tau_p \qquad (4.5)$$

where τ_p is added to account for path transmittance. The variables can be separated as shown in Eq. (4.5) based upon the following assumptions:

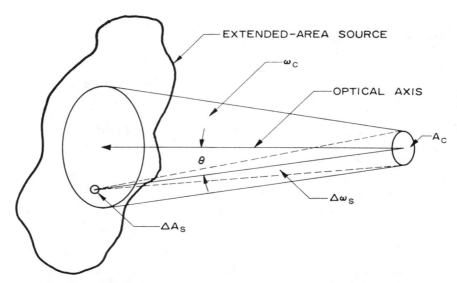

Figure 4.1   Geometry for transfer of flux from an extended-area source to a collector.

   1. The solid angle and the sterance [radiance] are independent of the area. The rule of thumb[4] of Sec. 3.2.4 applies here: provided the separation between the source and the entrance aperture is 10 to 20 times the maximum transverse dimension of the aperture or the source as defined by the sensor field of view,

$$\int_{A_s} dA_s \int_{\omega_s} \cos \theta \, d\omega_s = A_s \Omega_s \tag{4.6}$$

In this case, the solid angle Ω_s can be approximated by

$$\Omega_s = \frac{A_c}{s^2} \tag{4.7}$$

where s is the distance between the source and the entrance aperture A_c. (*Note:* The subscript on Ω corresponds to the location of the vertex of the cone that describes the solid angle.)

   2. The spectral sterance [radiance] is independent of area and wavelength over the spectral bandpass; thus

$$L(\lambda) \int_{\lambda} d\lambda = L(\lambda) \, \Delta\lambda = L \qquad [\text{W cm}^{-2} \, \text{sr}^{-1}] \tag{4.8}$$

when the bandpass, $\Delta\lambda$, is small.

   3. The path loss term τ_p, obtained using computer models as given above, can be included. Then the band flux is given by

$$\Phi = L\tau_p \frac{A_s A_c}{s^2} = L\tau_p \Upsilon \tag{4.9}$$

Equation (4.9) is often accurate enough for many calculations. Under these conditions, the invariance theorem leads to the following relationship for the throughput Υ:

$$\Upsilon = A_s \Omega_s = A_c \Omega_c \quad [\text{cm}^2 \text{ sr}] \tag{4.10}$$

where Ω_c is the solid angle subtended by the source A_s at the collector A_c. Thus, neglecting losses,

$$\Phi = LA_s \Omega_s = LA_c \Omega_c \quad [\text{W}] \tag{4.11}$$

This equation may be interpreted in the case of an extended-area source as follows: The flux in the beam is given by the product of the band sterance [radiance] L and the sensor throughput $A_c \Omega_c$. In this case it is convenient to consider the beam as the field of view of the sensor.

These considerations are illustrated for the case of the cloudy sky as a uniform source of diffuse radiation. For the first case consider a flat-plate receiver of diameter $d = 10$ cm whose field of view consists of the entire hemisphere.

Example 1: Find the flux on a flat-plate collector of diameter $d = 10$ cm from a cloudy sky.

Given: The average cloudy-sky band sterance [radiance] is $L = 2.40 \times 10^{-2}$ W cm^{-2} sr^{-1}.

Basic equations:

$$\Omega = \pi \sin^2 \Theta \quad [\text{sr}] \tag{3.6}$$

$$L = \frac{d^2 \Phi}{dA \cos \theta \, d\omega} \quad [\phi \text{ cm}^{-2} \text{ sr}^{-1}] \tag{3.19}$$

$$\Phi = L \int_A \int_\omega dA \cos \theta \, d\omega = L\Upsilon \quad [\phi] \tag{3.21}$$

Assumptions: The sky radiates uniformly over the entire hemisphere.

Note: The sterance [radiance] is defined to include path losses so the term τ_p is not appropriate here.

Solution: The throughput of the entrance aperture is given by

$$\Upsilon = A_c \pi \sin^2 \Theta = \frac{\pi d^2}{4} \pi \sin^2 90° = 246.7 \text{ cm}^2 \text{ sr}$$

where $\Theta = 90°$ for the hemisphere.
The total flux is given by

$$\Phi = L\Upsilon = 2.4 \times 10^{-2} [\text{W cm}^{-2} \text{ sr}^{-1}] \times 246.7 [\text{cm}^2 \text{ sr}] = 5.92 \text{ W} \quad \blacksquare$$

For a second case, suppose the sensor made use of a lens and appropriate field stop to provide for spatial resolution. Such a sensor could be used to measure the nonuniformity of an extended-area source such as the sky. In this case, not all the radiation falling upon the entrance aperture will pass through the field stop. The power that is effective in passing through the field stop can be calculated utilizing Eq. (3.6) again.

Example 2:   Find the effective flux entering the field stop.

   Given: The sterance [radiance] is given in Example 1. The half-angle field of view is $\Theta = 1°$. The lens diameter is 10 cm.

   Basic equations: Same as for Example 1.

   Assumptions: The sky radiates uniformly over the sensor field of view.

   Solution: The solid-angle field of view, Eq. (3.6), is

$$\Omega_c = \pi \sin^2 \Theta = 9.57 \times 10^{-4} \text{ sr}$$

The collector area is

$$A_c = \frac{\pi d^2}{4} = \frac{\pi \times 100}{4} = 78.54 \text{ cm}^2$$

The effective flux collected is

$$\Phi = LA_c\Omega_c = 2.4 \times 10^{-2} [\text{W cm}^{-2} \text{ sr}^{-1}] \times 78.54 [\text{cm}^2]$$
$$\times 9.57 \times 10^{-4} [\text{sr}]$$
$$= 1.8 \times 10^{-3} \text{ W} \qquad\qquad\qquad \blacksquare$$

4.3.2   Point Source

The pointance [intensity] is appropriate for sources that approximate a point source and is especially appropriate for sources for which the area is unresolvable, such as a star. The pointance can be derived from the definition for sterance [radiance] L as follows: The band flux incident upon a reference surface A_c from a source A_s as illustrated in Fig. 4.2 is given in terms of the sterance L by

$$\Phi = \int_{\omega_s} \int_{A_s} L_s \, dA_s \cos \theta \, d\omega_s \qquad [\phi] \qquad (4.12)$$

which can be written as

$$\Phi = \frac{L_s A_s A_c}{s^2} \qquad (4.13)$$

where the same assumptions are made as given above for the extended-area source.

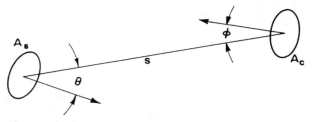

Figure 4.2   Power transfer from a source to a reference surface.

The average pointance [intensity] I is given by

$$I_s = \frac{\Phi}{\omega_s} = L_s A_s = L_s \pi r^2 \qquad (4.14)$$

where the projected source area is given by πr^2 and r is the source radius.
The average areance [irradiance] E is given by

$$E = \Phi/A_c = L_s A_s/s^2 = L_s \pi r^2/s^2 \qquad (4.15)$$

Using Eq. (4.14) yields

$$E = I_s/s^2 \qquad (4.16)$$

These considerations are illustrated for the case of the sun as a radiant point source. The throughput of a point source can be given by

$$\Upsilon = \int_\omega \int_A \cos\theta \, dA \, d\omega = \int_A \cos\theta \, dA \int_\omega d\omega = \pi r^2 \times 4\pi = 4\pi^2 r^2 \quad (4.17)$$

where r is the sphere radius, πr^2 is the projected area (the disk), and 4π is the solid angle of a sphere, or

$$\Upsilon = \int dA \int \cos d\omega = 4\pi r^2 \times \pi = 4\pi^2 r^2 \qquad (4.18)$$

where $4\pi r^2$ is the area of a sphere and π is the projected solid angle in any specific direction. The results are identical whether the projected area or the projected solid angle is used.

Example 3: Find the sterance (radiance) of the sun.

Given: The solar constant (measured) is 0.134 W/cm², neglecting atmospheric effects. The sun subtends a full angle of 0.532° at the earth.

Basic equations:

$$\Omega = \pi \sin^2\Theta \qquad [\text{sr}] \qquad (3.6)$$

$$E = d\phi/dA \qquad [\phi/\text{cm}^2] \qquad (3.34)$$

$$\Phi = \int_{\omega_s} \int_{A_s} L_s \, dA_s \cos\theta \, d\omega_s \qquad [\phi] \qquad (4.12)$$

Assumptions: The sun radiates as a uniform isotropic source.

Solution: For a point source, Eq. (4.12) can be written as

$$\Phi_c = L_s A_{ps} \omega_s = L_s A_{ps} A_c/s^2$$

where A_{ps} is the area of the solar disk and ω_s is the solid angle subtended by the collector at the sun. The solar constant

$$E_c = \Phi_c/A_c = L_s A_{ps}/s^2 = L_s \Omega_c \qquad [\text{W/cm}^2]$$

The quantity

$$\Omega_c = A_{ps}/s^2 = \pi \sin^2(\theta/2) = 6.77 \times 10^{-5} \qquad [\text{sr}]$$

Thus

$$L_s = \frac{E_c}{\Omega_c} = \frac{0.134[\text{W}/\text{cm}^2]}{6.77 \times 10^{-5}[\text{sr}]} = 1.979 \times 10^3 \text{ W cm}^{-2} \text{ sr}^{-1} \qquad \blacksquare$$

Example 4:   Find the total flux and the pointance [intensity] of the sun.

Given: Refer to Example 3. The sun-earth distance s is 1.5×10^{13} cm.

Basic equations:

$$\Phi = \int_{\omega_s} \int_{A_s} L_s \, dA_s \cos \theta \, d\omega_s \qquad [\phi] \qquad (4.12)$$

$$I_s = \Phi/\omega_s \qquad\qquad (4.14)$$

Assumptions: Same as for Example 3.

Solution: For an isotropic point source, Eq. (4.12) can be written as

$$\Phi = L_s A_{\text{ps}} 4\pi$$

since the sun radiates into 4π sr (a sphere). The radius of the solar disk is given by

$$r = s \sin(\theta/2) = 1.5 \times 10^{13}[\text{cm}] \sin(0.532/2) = 6.964 \times 10^{10} \text{ cm}$$

and the projected area of the solar disk is

$$A_{\text{ps}} = \pi r^2 = \pi (6.946 \times 10^{10})^2 = 1.5235 \times 10^{22}$$

The total flux is therefore

$$\Phi_s = L_s A_{\text{ps}} \times 4\pi = 1.979 \times 10^3 [\text{W cm}^{-2} \text{ sr}^{-1}] \times 1.5235 \times 10^{22} \times 4\pi$$
$$= 3.789 \times 10^{26} \text{ W}$$

The average pointance [intensity] is

$$I_s = \frac{\Phi_s}{4\pi} = \frac{3.789 \times 10^{26}[\text{W}]}{4\pi[\text{sr}]} = 3.015 \times 10^{25} \text{ W}/\text{sr} \qquad \blacksquare$$

4.4   PROPAGATION OF FLUX IN A PLASMA OR GAS

The emission, scattering, and absorption in a plasma or an optically thin gas is of interest in atmospheric studies or in taking into account the modification flux undergoes when passing through an optically thin gas.[5]

Emission within the gas is described using the entity of the photon volume emission rate $\rho(s)$ with units of quanta per second per cubic centimeter (q s^{-1} cm^{-3}). If we assume isotropic radiation, the effective incremental photon flux $\Delta\phi_p$ incident upon a sensor aperture is given (see Fig. 4.3) by

$$\Delta\Phi_p = \tau_p(s) \frac{\rho(s)}{4\pi} \Delta V \cos\theta \frac{A_c}{s^2} \qquad [\text{q}/\text{s}] \qquad (4.19)$$

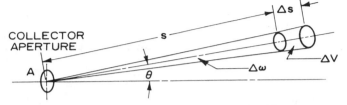

Figure 4.3 Geometry for radiation in an optically thin gas.

where ΔV is the incremental radiating volume, s is the distance from the aperture A_c to ΔV, and the transmissivity $\tau_p(s)$ takes into account losses due to scattering and absorption.

The incremental volume can be expressed in spherical coordinates as

$$\Delta V = s^2 \Delta s \, \Delta \omega \qquad [\text{cm}^3] \qquad (4.20)$$

where $\Delta \omega$ is the solid angle of the incremental volume subtended at the sensor aperture. Thus, the total flux incident upon the aperture is given by

$$\Phi_p = \frac{A_c}{4\pi} \int_\omega \left[\int_s \rho(s)\tau_p(s) \, ds \right] \cos\theta \, d\omega \qquad [\text{q/s}] \qquad (4.21)$$

where

$$\int_\omega \cos\theta \, d\omega = \Omega_c \qquad [\text{sr}] \qquad (4.22)$$

Thus

$$L_p = \frac{\Phi_p}{A_c \Omega_c} = \frac{1}{4\pi} \int_s \rho(s)\tau_p(s) \, ds \qquad [\text{q s}^{-1}\text{cm}^{-2}\text{sr}^{-1}] \qquad (4.23)$$

which gives the photon sterance L_p as though the gas were radiating as a surface.

Chamberlain[6] states that the emission rate integrated over the whole column (path) is given by

$$4\pi L_p = \int_s \rho(s) \, ds \qquad (4.24)$$

assuming no losses. Since these assumptions are generally not satisfied, $4\pi L_p$ represents an *apparent* emission rate. For this reason, and to have a more convenient unit, the rayleigh (R) was introduced. If L_p is measured in units of 10^6 q s^{-1}, then $4\pi L_p$ is in rayleighs.

1 Rayleigh = apparent emission rate of 10^6 q s^{-1}cm^{-2} (column)

The word "column" signifies that it is an integrated quantity.

Chamberlain refers to $4\pi \times 10^6 L_p$ as the apparent volume emission rate, and its utility becomes apparent since the derivative of the photon sterance [radiance] $L_p(s)$, measured along a path s, yields the distribution of the volume emission rate along the same path. That is,

$$4\pi \frac{dL_p(s)}{ds} = \rho(s) \qquad [\text{q s}^{-1}\text{cm}^{-3}] \qquad (4.25)$$

This applies in the case where a sensor mounted upon a rocket platform measures the total overhead sterance [radiance] as a function of height as it passes through the atmosphere. Equation (4.21) yields the vertical distribution of the volume emission rate.

The rayleigh is obtained from the radiant sterance [radiance] by[7,8]

$$1 \text{ Rayleigh} = 2\pi\lambda \times 10^{13} L_e \quad [R] \tag{4.26}$$

where λ is entered directly in μm and L_e has the units W cm^{-2} sr^{-1}.

4.5   FIBER OPTICS LINK

The attenuation of a fiber optic link in a communication system is of interest. Current developmental efforts are directed toward reducing the losses in the fiber optics channel due to attenuation of the optical signal to extend the range between repeater amplifiers.[9]

Example 5:   Find the distance between repeater amplifiers for a fiber optics link.

Given: The source emits 430 mW of peak pulsed power, the fiber link exhibits 10 dB/km attenuation, the detector NEP $= 10^{-5}$ W at the required bandwidth, and a minimum SNR of 10 is required. Neglect coupler losses.

Basic equation:

$$dB = 10 \log(\Phi_o/\Phi_i)$$

Assumptions: Neglect coupler losses.

Solution: The SNR of 10 requires that 10^{-4} W be incident upon the detector. The allowed link loss is given as the ratio of source output power, Φ_o to the power incident upon the detector, Φ_i, as

$$\Phi_o/\Phi_i = 0.43/10^{-4} = 4300$$

The fiber link loss expressed in decibels, as a function of the power ratio, is

$$10 \log(\Phi_o/\Phi_i) = 36.33 \text{ dB}$$

Thus, a 3.6-km link is possible between repeater amplifiers.   ∎

EXERCISES

1. Given: The sterance of the sun is $L(\text{sun}) = 1.98 \times 10^3$ W cm^{-2} sr^{-1}, the sun subtends a full angle of 0.53° at the earth, and the collector size is 4×8 ft. Assume the collector is normal to the sun's rays. Find the total power incident upon a flat-plate collector from the sun on a clear day.

2. Given: The sky sterance is $L(\text{sky}) = 2.40 \times 10^{-2}$ W cm^{-2} sr^{-1}. Assume the entire sky (hemisphere) is a uniform diffuse extended-area source. Find the power collected on a cloudy day (refer to Exercise 1).

3. A 36-in. diameter engine exhaust radiates as a diffuse source. The source areance [exitance] $M = 1.8 \times 10^2$ W/cm^2. Calculate the areance [irradiance] observed at a range of 100 yd normal to the surface.

4. A spy satellite records a bright flash of light near the surface of a hostile country. Peak areance [irradiance] observed at a range of 350 km is 1.8×10^2 W/cm^2. Find the total energy rate of the source. Assume the source radiates isotropically.

5. Find the power required to transmit 10 km between repeater amplifiers on a fiber link. Given that the attenuation is 2 dB/km and the incident power required at the receiver (detector) is 10^{-4} W.

6. A photometer measured the photon sterance emitted by an optically thin atmosphere at $L_p = 1.6 \times 10^{-10}$ q s^{-1} cm^{-2} sr^{-1} over a path length of 2 km. Find the volume emission rate ρ in q s^{-1} cm^{-3}. Assume the volume emission rate is uniform throughout the atmosphere and the transmittance $\tau(s)$ is unity.

7. An extended-area source radiates $L_s = 2.3 \times 10^{-3}$ W cm^{-2} sr^{-1}. A sensor has a 2-cm diameter collector and a half-angle field of view of 1°. Find the flux incident upon the collector within the field of view.

8. A distant small-area source radiates $I_s = 2.8 \times 10^2$ W/sr. The source of diameter 200 cm is at a distance of 2 km from a sensor aperture of 2-cm diameter. Find the flux incident upon the aperture.

9. Given that a source exhibits a band sterance [radiance] of 1.27×10^{11} W cm^{-2} sr^{-1} at 6600 cm^{-1}, find the photon areance [exitance] in terms of rayleighs.

REFERENCES

1. E. E. Bell, "Radiometric Quantities, Symbols, and Units," *Proc. IRE*, **47**, 1432–1434 (1959).
2. W. L. Wolfe and G. J. Zissis, Eds., *The Infrared Handbook*, Office of Naval Research, Dept. of the Navy, Washington, DC, 1978, p. 5-3.
3. F. E. Nicodemus, Ed., "Self-Study Manual on Optical Radiation Measurements. Part 1—Concepts," *Natl. Bur. Stand. (U.S.), Tech. Note* No. 910-2, 1978, p. 31.
4. *Ibid.*, p. 34.
5. *Ibid.*, p. 41.
6. J. W. Chamberlain, *Physics of the Aurora and Airglow*, Academic Press, N.Y., 1961, p. 569.
7. D. J. Baker, "Rayleigh, the Unit for Light Radiance," *Appl. Opt.*, **13**, 2160–2163 (1974).
8. D. J. Baker and G. J. Romick, "The Rayleigh: Interpretation of the Unit in Terms of Column Emission Rate or Apparent Radiance Expressed in SI Units, *Applied Optics*, **15**, 1966–1968 (1976).
9. D. L. Baldwin et al., "Optical Fiber Transmission System Demonstration over 32 km with Repeaters Data Rate Transparent up to 2.3 Mbits/s," *IEEE Trans. Comm.*, **Com-26**, 1045–1055 (1978).

chapter 5

The Optical Subsystem

5.1 INTRODUCTION

The previous chapters developed the geometrical parameters and path-loss factors of flux transfer. The objective of this chapter is to consider the selective collection of flux by an optical subsystem where in each case the *effective* flux incident upon the detector is obtained by integration over the appropriate variables.

This chapter emphasizes the parameters of the optical subsystem, which are described in terms of general parameters or *figures of merit* that relate to its flux-gathering capability, which can be utilized to write the radiometric performance equation.

The optical subsystem design determines the spatial, spectral, and polarization properties of the sensor. The design criteria for the optical subsystem are best established in terms of applications considerations as follows.

The objective of the design of the *spatial response* is to collect the radiant flux from a specific region referred to as the "field of view." The field of view should be matched to the geometry of the radiant source to maximize source flux collected and minimized unwanted flux.

There are two general classifications of application goals:

1. Maximize flux collected over a spatial region.
2. Maximize the resolving power within a spatial region.

The first has application in the detection of flux from an extended-area source. In the ideal, such a design provides a measure of the total integrated flux over the specified region. The second has application in mapping the variations of the flux within a scene. The spatial resolution determines the detail to which the variations within the scene can be resolved. The spatial resolution is usually characterized in terms of the instantaneous field of view in degrees, radians, or steradians. The resolution must also be maximized in the case of the detection of flux from a point source in which the effects of the background must be reduced.

The objective of the design of the *spectral response* is to collect the radiant energy from a specific spectral region referred to as the "spectral band." The spectral response should be matched to the spectral distribution of the radiant source to maximize source flux collected and minimize unwanted flux. The spectral domain is mathematically similar to the spatial domain except that it involves a single dimension (wavelength or wavenumber), whereas the spatial domain involves two or sometimes three dimensions.

Here too there are two general classifications of application goals:

1. Maximize the flux collected over the spectral band.
2. Maximize the resolving power within a spectral band or region.

The first has application in the detection of flux from a source that radiates over a spectral band. In the ideal, such a design provides a measure of the total integrated flux within the specified region. The second has application in mapping the variations of the flux in terms of its spectral power density function, which is usually referred to as the *spectrum*. It may also have application in the detection of a nearly monochromatic source in which the effect of neighboring emissions must be reduced. The spectral resolution is generally characterized in terms of *resolving power*, which is defined as the ratio of the bandwidth to the center wavelength or wavenumber.

The objective of the design of the *polarization response* is to provide for the characterization of the source flux polarization or to avoid unwanted metrological effects. Polarization is a fundamental property of electromagnetic radiation. The change that takes place in the polarization characteristics of electromagnetic radiation, as it interacts with materials and with electric and magnetic fields, is a convenient and accurate diagnostic tool.[1]

5.2 IDEAL AND NONIDEAL FIELDS OF VIEW

The throughput of a sensor was given in Chap. 4 as the product of the entrance aperture area A_c [cm^2] and the projected solid-angle field of view Ω_c [sr]. The projected solid angle, for a circularly symmetric field of view, was obtained by integration as $\Omega_c = \pi \sin^2 \Theta$, where Θ is the half-angle field of view in degrees.

The field of view of the sensor must be matched to the source. Thus the objective in the design of the spatial response is to collect the total flux in the specific region referred to as the field of view. The total flux is given by

$$\Phi_T = \int_\omega \Phi(\theta, \phi)\, d\Omega \qquad [\phi] \qquad (5.1)$$

where the projected solid angle Ω is given by

$$\Omega = \int_\omega \cos\theta\, d\omega \qquad [\text{sr}] \qquad (5.2)$$

The sensor output V is given by

$$V = \mathscr{R}_0 \int_{\text{hem}} \mathscr{R}(\theta, \phi)\Phi(\theta, \phi)\, d\Omega \qquad [\text{V}] \qquad (5.3)$$

where the integration is carried out over the entire hemisphere, $\mathscr{R}_0$ is the peak responsivity in volts per unit of flux, V/ϕ, along the optical axis, and $\mathscr{R}(\theta, \phi)$ is the normalized spatial response (the field of view).

From this, the measured source flux is given by

$$\Phi_m = \frac{V}{\mathscr{R}_0} = \int_{\text{hem}} \mathscr{R}(\theta, \phi)\Phi(\theta, \phi)\, d\Omega \qquad [\phi] \qquad (5.4)$$

which, by Eq. (5.1), is equal to Φ_T only if

$$\int_{\text{hem}} \mathscr{R}(\theta, \phi)\Phi(\theta, \phi)\, d\Omega = \int_\Omega \Phi(\theta, \phi)\, d\Omega \qquad [\phi] \qquad (5.5)$$

Unfortunately, the equality in Eq. (5.5) is valid only under two special conditions, neither of which is perfectly realized in practice.

1. The equality in Eq. (5.5) is valid for an *ideal sensor* relative spatial response function, $\mathscr{R}(\theta, \phi)$, which is defined as one that has unity response over the sensor field of view Ω and is zero elsewhere. Then the term $\mathscr{R}(\theta, \phi)$ is a unity constant that passes through the integral and Eq. (5.5) is identically true.
2. The equality in Eq. (5.5) is also valid for the special case where the spatial flux $\Phi(\theta, \phi) = \Phi_0$ is a constant that passes through the integral. This is realized for a point source or a uniform extended-area source.

The ideal sensor, case 1, does not exist; therefore, Eq. (5.4) yields the true flux only when the source is a point source or a spatially uniform extended-area source. Corrections can be applied for nonuniform spatial distributions[2]; however, in the absence of such corrections the flux should be reported as "peak normalized."[3] A good deal of the design effort in

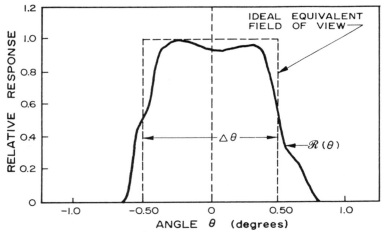

Figure 5.1 Field-of-view representation as a cross section.

electro-optical instrumentation, is expended in an attempt to develop a nearly ideal field of view to reduce the errors that result from nonuniform sources.

Under the conditions of a point source or a uniform source, Eq. (5.5) becomes

$$\int_{\text{hem}} \mathscr{R}(\theta, \phi) \, d\Omega = \int_{\Omega} d\Omega = \Omega \qquad [\text{sr}] \qquad (5.6)$$

and indicates that the integral of the nonideal response function must be equal to that of the ideal response function. The field-of-view response can be considered in terms of a single dimension $\mathscr{R}(\theta)$ for a circularly symmetrical field of view and can be represented as a cross section as illustrated in Fig. 5.1.

Equation (5.6) yields the area under the $\mathscr{R}(\theta)$ curve, which is exactly equal to the full angle $\Delta\theta$ (see Fig. 5.1) of the equivalent ideal field of view. Thus for practical systems, the field of view is defined in terms of the angles that give the edges of the equivalent ideal response function.

5.3 IDEAL AND NONIDEAL SPECTRAL BANDWIDTHS

Mathematical similarity exists between the spatial and spectral domains except for the terms and the fact that the field of view is a two-dimensional entity.

The spectral properties of the sensor must also be matched to the source. Thus the objective in the design of the spectral response is to measure the total flux in a specific spectral band $\lambda_2 - \lambda_1$. The total flux is given by

$$\Phi_T = \int_{\lambda_1}^{\lambda_2} \Phi(\lambda)\, d\lambda \qquad [\phi] \qquad (5.7)$$

The sensor output V is given by

$$V = \mathscr{R}_0 \int_0^\infty \mathscr{R}(\lambda)\Phi(\lambda)\, d\lambda \qquad [V] \qquad (5.8)$$

where $\mathscr{R}_0$ is the absolute responsivity with units of volts per unit flux (V/ϕ), $\mathscr{R}(\lambda)$ is the normalized relative spectral response, and the integration is over all wavelengths.

The measured source flux is given by

$$\Phi_m = V/\mathscr{R}_0 = \int \mathscr{R}(\lambda)\Phi(\lambda)\, d\lambda \qquad [\phi] \qquad (5.9)$$

which, by Eq. (5.7), is equal to Φ_T only if

$$\int_0^\infty \mathscr{R}(\lambda)\Phi(\lambda)\, d\lambda = \int_{\lambda_1}^{\lambda_2} \Phi(\lambda)\, d\lambda \qquad (5.10)$$

Unfortunately, the equality in Eq. (5.10) is valid under only two special conditions, neither of which is perfectly realized in practice:

1. The equality in Eq. (5.10) is valid for an ideal sensor relative spectral response function, $\mathscr{R}(\lambda)$, which is defined as one that has unity response over the range λ_1 to λ_2 and is zero elsewhere. Then the term $\mathscr{R}(\lambda)$ is a unity constant that passes through the integral, and Eq. (5.10) is identically true.
2. The equality in Eq. (5.10) is also valid for the special case where the spectral flux $\Phi(\lambda) = \Phi_0$, a constant that passes through the integral. This is realized for a monochromatic or a uniform spectral source.

The ideal sensor, case 1, does not exist; therefore, Eq. (5.9) yields the true flux only when the source is either monochromatic or spectrally uniform. Corrections can be applied for nonuniform spectral densities[4]; however, in the absence of such corrections the flux should be reported as "peak normalized." A good deal of the design effort in electro-optical instrumentation is expended in an attempt to develop a nearly ideal spectral bandpass to reduce the errors that result from nonuniform sources.

Under the conditions of a monochromatic or a uniform source, Eq. (5.10) becomes

$$\int_0^\infty \mathscr{R}(\lambda) = \int_{\lambda_1}^{\lambda_2} d\lambda = \lambda_2 - \lambda_1 = \Delta\lambda \qquad [\mu m] \qquad (5.11)$$

which indicates that the nonideal response function must be equivalent in

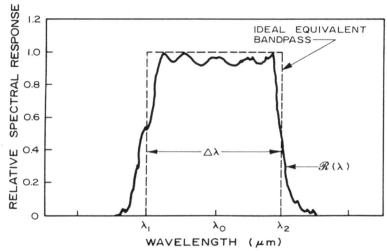

Figure 5.2 Illustration of a practical bandpass filter and its equivalent-area ideal representation.

area to that of the ideal response function. Figure 5.2 illustrates a typical peak-normalized spectral-bandpass curve. For practical systems, the bandpass is defined in terms of the wavelengths λ_1 and λ_2 that give the edges of the equivalent (area) ideal square response function or in terms of $\Delta\lambda$, the ideal equivalent bandwidth.

5.4 POLARIZATION RESPONSE

The state of polarization of a beam of radiant flux can be completely specified in terms of the Stokes' vector,[5,6] which consists of a set of four entities called Stokes' parameters.

Serious problems arise from undesired polarization properties of certain optical sensors. This is because the sensor response is often a function of the polarization of the incident flux. All natural radiation is polarized to some extent and can be a problem because sensor response may depend upon the physical orientation of the sensor relative to the source.

Any optical sensor that utilizes mirrors, slits, gratings, or beam splitters may exhibit some degree of polarization sensitivity.

Sensor designs that utilize refractive lenses and antireflection coatings, mirrors and high reflectance coatings, or thin-film spectral filters are generally not polarization-sensitive. This is because the optical components are usually circularly symmetrical, and the radiation falls at or near normal incidence upon these optical components.[7]

5.5   EFFECTIVE FLUX

The effectiveness of the flux in evoking a response in the detector at any wavelength λ is the product of the source flux $\Phi(\lambda)$ and the relative spectral response $\mathscr{R}(\lambda)$. For example,

$$\Phi_{\text{eff}} = \int \mathscr{R}(\lambda)\Phi(\lambda)\, d\lambda \qquad (5.12)$$

for a radiometer, and

$$\Phi_{\text{eff}}(\lambda) = \frac{1}{\Delta\lambda} \int \mathscr{R}(\lambda)\Phi(\lambda)\, d\lambda \qquad (5.13)$$

for a spectrometer, where $\Delta\lambda$ is the equivalent ideal spectral bandwidth. Generally, for high-resolution spectrometers, $\Phi(\lambda)$ can be considered constant over $\Delta\lambda$. Then, by Eq. (5.13),

$$\Phi_{\text{eff}}(\lambda) = \frac{\Phi(\lambda)}{\Delta\lambda} \int \mathscr{R}(\lambda)\, d\lambda = \Phi(\lambda) \qquad (5.14)$$

The magnitude of the effective flux calculated using Eq. (5.12) depends upon how $\mathscr{R}(\lambda)$ is normalized. Ordinarily, the relative response is peak-normalized and the effective flux is termed "peak-normalized flux."[3]

Even for wideband radiometers, the normalized flux can often be approximated with sufficient accuracy for a feasibility study by assuming that $\Phi(\lambda) = \Phi(\lambda_0)$, a constant, over the spectral bandwidth $\Delta\lambda$. Then

$$\Phi_{\text{eff}} = \Phi(\lambda_0)\,\Delta\lambda \qquad (5.15)$$

where λ_0 is the center wavelength and $\Delta\lambda$ is the equivalent (area) ideal half-power bandwidth.

5.6   RELATIVE APERTURE OR f-NUMBER

The throughput of a beam of radiant flux was given in Sec. 3.2.4. There it was pointed out that the throughput is invariant in a homogeneous medium and applies to a beam of flux, to the entrance aperture of a sensor, and to detectors. Thus, the throughput provides a figure of merit for the optical subsystem.

The *relative aperture* or *f-number*,[8] like the throughput, is a measure of the "flux-gathering power" of the optical system. The f-number, F, is given by the ratio of the effective focal length f to the entrance aperture diameter:

$$F = f/D \qquad (5.16)$$

as illustrated in Fig. 5.3.

The f-number is also obtained from Fig. 5.3 as

$$F = \frac{1}{2\eta \sin\alpha} \qquad (5.17)$$

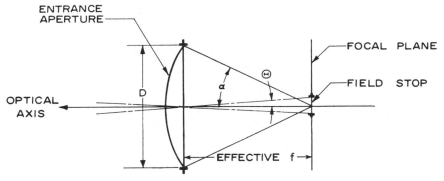

Figure 5.3 Optical schematic for a simple optical system containing a field stop in the focal plane of a collector lens.

where η is the index of refraction, which for air is approximately unity. The quantity $\eta \sin \alpha$ is referred to as the *numerical aperture*. The limiting value for F occurs for $\alpha = 90°$ (see Fig. 5.3), which yields $F = 1/2\eta$.

The throughput of the field stop of Fig. 5.3 is given as the product of the stop area A_{fs} and the projected solid angle of the lens cone Ω_{fs}. Note that the rays describing the field-of-view cone pass through the center of the lens undeviated. The projected solid angle is given in terms of α by

$$\Omega_{fs} = \pi \sin^2\alpha \quad [\text{sr}] \qquad (5.18)$$

and is obtained in terms of F by taking the ratio of the lens projected area to the effective focal length squared:

$$\Omega_{fs} = \pi D^2/4f^2 = \pi/4F^2 \qquad (5.19)$$

The invariance of throughput is illustrated for the simple optical system of Fig. 5.3 as follows: The throughput of the field stop is given by

$$A_{fs}\Omega_{fs} = \text{stop area} \times \frac{\text{lens area}}{f^2} \quad [\text{cm}^2 \, \text{sr}] \qquad (5.20)$$

and the throughput of the lens is given by

$$A_L\Omega_L = \text{lens area} \times \frac{\text{stop area}}{f^2} \qquad (5.21)$$

Equations (5.20) and (5.21) are identical.

Although the invariance theorem holds, the throughput calculated using Eqs. (5.20) and (5.21) may be in error because the focal length is not large compared with the linear dimensions of the lens and stop.

The throughput, for low *f*-numbers, must be calculated using configuration factors as given in Sec. 3.2.5. However, the accuracy of Eqs. (5.20) and (5.21) is sufficient for most feasibility studies.

Example 1:   Find the exact throughput for the $F = 1$ lens and field stop of
Fig. 5.3.

Given: Lens diameter $= 10$ cm, effective focal length $= 10$ cm, and the
circular stop diameter is 0.2 cm.

Basic equations: Approximate solution:

$$A_{fs}\Omega_{fs} = \text{stop area} \times \frac{\text{lens area}}{f^2} \qquad [\text{cm}^2 \text{ sr}] \qquad (5.20)$$

$$A_L\Omega_L = \text{lens area} \times \frac{\text{stop area}}{f^2} \qquad\qquad (5.21)$$

Exact solution:

$$\Upsilon_1 = F_1\pi A_1 \qquad\qquad (3.13)$$

$$F_1 = \tfrac{1}{2}\left(z - \sqrt{z^2 - 4x^2y^2}\right) \qquad\qquad (3.14)$$

$$x = r_2/s, \qquad y = s/r_1, \qquad z = 1 + (1 + x^2)y^2 \quad (3.15)$$

Solution: Exact solution for throughput of stop:

$x = r_2/s = 5/10 = 0.5$
$y = s/r_1 = 10/0.1 = 100$
$z = 1 + (1 + x^2)y^2 = 1 + \left[1 + (0.5)^2\right] \times 100^2 = 1.2501 \times 10^4$
$F_1 = \tfrac{1}{2}\left(z - \sqrt{z^2 - 4x^2y^2}\right) = 1.99985 \times 10^4$
$\Upsilon_{fs} = F_1\pi A_1 = 1.97377 \times 10^{-2} \qquad [\text{cm}^2 \text{ sr}]$

Exact solution for throughput of lens:

$x = r_2/s = 0.1/10 = 0.01$
$y = s/r_1 = 10/5 = 2$
$z = 1 + (1 + x^2)y^2 = 1 + \left[1 + (0.01)^2\right] \times 2^2 = 5.0004$
$F_2 = \tfrac{1}{2}\left(z - \sqrt{z^2 - 4x^2y^2}\right) = 7.995 \times 10^{-5}$
$\Upsilon_L = F_2\pi A_2 = 1.97379 \times 10^{-2}$

Approximate solution for throughput of lens or field stop:

$$A_{fs}\Omega_{fs} = \text{stop area} \times \frac{\text{lens area}}{f^2} = \frac{\pi(0.2)^2}{4} \times \frac{\pi \times 10^2}{4} \times \frac{1}{10^2}$$

$$= 2.4674 \times 10^{-2}$$

Note: The configuration factors of Sec. 3.2.5 yield essentially identical
solutions for the throughput of the lens and the field stop of 1.9738×10^{-2} cm^2 sr. The approximate solutions given in Eqs. (5.20) and (5.21)
also yield essentially identical values for the throughput of the lens and

the stop of 2.4674×10^{-2} cm^2 sr. Thus, the invariance theorem holds for either solution. However, the exact solution yields values of throughput that are 25% lower than the approximate values for an $F = 1$ system. ∎

Any optical system, regardless of how many elements are used or how complex it is, can usually be represented by a simple equivalent system like that of Fig. 5.3 for throughput considerations, where the field stop is the detector.

5.7 OPTICAL CHOPPER LOSSES

An optical chopper is a mechanical device, such as a rotating sector wheel, that is used in optical systems to periodically interrupt the flux incident upon the detector. The resultant electric signal can be amplified using an ac-coupled system. The primary reason for using an optical chopper is to transform the optical signal information band to a frequency for which the detector noise characteristics are more favorable.[9] Such a technique avoids problems of $1/f$ noise and low-frequency drift. It also encodes the radiant-source flux to provide for discrimination between the source flux and various possible background sources of flux.

The chopper functions as a radiometric standard, since the ac circuits measure the *difference* between the source flux and the chopper flux. Often the chopper is designed to yield a flux relatively near zero; this can be achieved for IR systems by cooling the chopper to reduce emissions or for short-wavelength systems by adequate baffling to reduce reflected flux. In this case the electric output signal is proportional to the absolute flux.

A *reflective* chopper is used in some systems that are designed to direct the flux from an internal radiometric standard to the detector. The electric output signal is proportional to the difference between the source flux and the nonzero reference flux. The reference flux may be dynamically adjusted to produce an output *null*; in this case the source flux is equal to the reference flux. Such a system avoids problems of variations in detector sensitivity and nonlinear response.

The interruption of the source flux with an optical chopper reduces the total source flux reaching the detector. There is an optical loss that depends upon the nature of the convolution of the chopper with the flux field. The optical loss also depends to some extent upon the ability of the signal-conditioning electronics to reproduce the waveform produced by the detector in response to the chopped flux.

The *chopping factor* β is a number greater than 1 that represents the optical loss introduced by the chopper. It is given by the ratio of the minimum detectable flux ϕ_m to the detector noise equivalent flux (NEF):

$$\beta = \phi_m / \text{NEF} \qquad (5.22)$$

Equation (5.22) can be expressed by

$$\beta = \left(\frac{1}{T} \int_0^{T/2} f(t)\, dT - \frac{1}{T} \int_{T/2}^T f(t)\, dt \right)^{-1} \tag{5.23}$$

where T is the period of the optical waveform and $f(t)$ is the relative shape of the waveform over one period. The optimum, or ideal, design is one in which the chopper produces a symmetrical square wave for which the denominator of Eq. (5.23) is $1/2$ and $\beta = 2$. These considerations lead to the conclusion that for a chopped sensor the optimum design is only half as sensitive as the ideal (noiseless) dc-coupled sensor.[10]

The requirement for abrupt symmetrical chopping can be supported by an intuitive argument: The sensor must make an estimate of the levels of the source flux and the chopper reference flux and compare them. Since the quality of the estimate increases with signal-to-noise ratio, it follows that (1) abrupt chopping provides the maximum average flux for the estimate and (2) the accuracy of the measurement is no better than the poorer estimate of either the source flux or the reference flux, so they must be estimated equally well, which implies equal time or symmetrical chopping.

The optimum design (square-wave sensor) with $\beta = 2$ is difficult to achieve in practice. There are two nonoptimum designs of note: The first, a sine-wave design, results from the use of tuned signal-conditioning amplifiers. The second, a triangular-wave sensor, results from using a chopper and flux field of approximately the same size.

The tuned amplifier selects the fundamental component of the square-wave signal. The value of the denominator of Eq. (5.23) is found by Fourier analysis to be $1/2.46$. The triangular-wave sensor results from using a chopper sector approximately equal in width to the radiant field. The value of the denominator of Eq. (5.23) for a triangular wave is $1/4$.

Thus the chopping factor can vary from 2 for the ideal to 4 for the triangular-wave sensor. Practical choppers are limited in size, so that square-wave chopping is only approximated. Tuned amplifiers are often used to reduce stray noise pickup; thus, a chopping factor of 3 is representative of many systems and is a useful value for feasibility studies.

For tuned systems the chopping factor accounts for the conversion from peak flux to rms flux.

5.8 REFLECTANCE AND TRANSMITTANCE LOSSES

Optical losses occur in the optical system because of nonideal reflectance and transmittance of flux by mirrors, lenses, and windows. These losses are expressed in accordance with the definitions for reflectance, absorptance, and transmittance; see Eq. (4.1). The net effect of losses in an optical system is obtained by taking the product of the transmittance and/or reflectance of all the components in the system. The symbol τ_e is used to express the total effect in terms of "optical efficiency."

Losses in refractive and reflective systems can be reduced substantially through the use of optical coatings.

5.9 OPTICAL SYSTEM IMAGE QUALITY

The quality of an optical system is usually expressed in terms of resolution or image size for a point source. The image of a point source is not a point; its size is determined by (1) the size of the system aperture because of diffraction effects and (2) the *f*-number for spherical aberrations. The image of a point source is referred to as the *blur*, and its size can be specified in angular measure or in terms of its linear diameter. The angular blur (in radians) is given by the ratio of the blur diameter to the effective focal length.

The most fundamental limitation to aspheric optical image quality is that resulting from diffraction effects. Generally, aberrations and surface imperfections can be reduced by good design to approach diffraction-limited conditions.

The blur circle for a perfect optical system, one with no aberrations or other physical imperfections, is called the "airy disk"; and takes the form of a central blur of light surrounded by alternating light and dark rings of rapidly decreasing intensity. The central blur contains 84% of the energy, so the diameter of the first dark ring about this central disk is a convenient measure of the size of the blur circle.[11] The angular size of the blur is given by

$$\theta = \frac{2.44\lambda}{D} \quad [\text{rad}] \qquad (5.24)$$

where D is the effective aperture (entrance pupil diameter) of the system (see Fig. 5.3), and the diameter of the blur circle is given by

$$d = 2.44\lambda F \qquad (5.25)$$

where λ is the wavelength and F is the relative aperture (or *f*-number). The diameters D in Eq. (5.24) and d in (5.25) must have the same dimensions as λ, and θ is given in radians. Spherical aberrations are given in terms of formulas and graphic representations of spherical aberrations of simple refractive lenses and spherical mirrors in the *Infrared Handbook*.[12]

5.10 FLUX INCIDENT UPON THE DETECTOR

To this point in the text, methods have been presented to characterize sources of optical flux and to calculate the quantity of flux emitted or reflected into a beam. The optical throughput of a beam has been defined and shown to be invariant. Furthermore, it has been shown that the invariance of throughput applies to optical systems, which makes it possible to calculate the effective flux incident upon the detector. Finally, optical losses due to absorption and scattering in the intervening medium, chopping, and the nonideal reflectance and transmittance of optical components have been described. It is now appropriate to evaluate the effective flux incident upon the detector.

The effective flux incident upon a detector is given in general by

$$\Phi_{\text{eff}} = \tau_p \tau_e \int_\lambda \int_\theta \int_\phi \Phi_s(\lambda, \theta, \phi) \mathscr{R}(\lambda) \, d\lambda \, d\theta \, d\phi / \beta \qquad (5.26)$$

where the source flux Φ_s is a function of wavelength and the geometric properties θ and ϕ of the radiated energy.

Equation (5.26) can be written in a more manageable form for a specific radiant entity. For example, the effective flux incident upon the detector is given for a sterance [radiance] source as follows:

$$\Phi_{\text{eff}} = \tau_p \tau_e \int_{A_s} \int_{\omega_s} \int_\lambda L_s(\lambda) \mathscr{R}(\lambda) \cos \theta \, dA_s \, d\omega_s \, d\lambda / \beta \qquad (5.27)$$

where τ_p is the path transmittance.

Equation (5.27) can be integrated, as given in Sec. 4.3.1, when the variables are independent, as

$$\Phi_{\text{eff}} = \frac{L_s(\lambda) \, \Delta\lambda \, A_s \Omega_s \tau_e \tau_p}{\beta} = \frac{L_s(\lambda) \, \Delta\lambda \, \tau_e \tau_p A_{\text{fs}} \pi}{4\beta F^2} \qquad (5.28)$$

where $\Delta\lambda$ is the equivalent ideal spectral bandwidth given by Eq. (5.11), $A_s \Omega_s$ is the equivalent ideal throughput of the beam (assuming the field of view of the sensor is matched to the spatial properties of the source), and $A_{\text{fs}} \pi / 4F^2$ [see Eq. (5.19)] is the throughput of the sensor field stop for the simple optical system of Fig. 5.3.

The right-hand term of Eq. (5.28) can be interpreted in terms of maximizing the flux incident upon the detector as follows: The throughput must be maximized to maximize Φ. This is accomplished, at least in principle, independently of the sensor field of view, by maximizing the field-stop area and minimizing the f-number, F. As indicated in Sec. 5.6, the simple equivalent representation of the optical system of Fig. 5.3 includes the detector as the stop.

There exist physical limits for which detectors can be fabricated. In addition, the properties of many detectors are such that the noise increases with area. So the signal-to-noise ratio may not be increased simply by increasing A_{fs}. These factors are considered in detail in the following chapters.

The f-number, F, is the most significant parameter in Eq. (5.28). The flux and the signal-to-noise ratio are inversely proportional to the f-number squared. The smaller the f-number, the greater the flux-collecting ability of the system. The cost of a system, as well as the throughput, is probably inversely related to F^2. However, the ultimate limit to the smallness of the f-number is set by theory at $F = 0.5$ for air.

Similarly, the effective flux incident upon the detector from a point source is given by

$$\phi_{\text{eff}} = \frac{I_s(\lambda) \, \Delta\lambda \, \omega_s \tau_p \tau_e}{\beta} = \frac{I_s(\lambda) \, \Delta\lambda \, A_c \tau_p \tau_e}{\beta s^2} \qquad (5.29)$$

The solid angle ω_s is written in terms of A_c/s^2 as the angle into which the flux is being collected by A_c, the entrance aperture, over the distance s. Thus we could conclude that the entrance aperture area should be enlarged without limit; this implies a field of view approaching zero. However, there are practical and theoretical limitations to how small the field of view can be made.

Given that the field of view is fixed at some value, increasing A_c is equivalent to reducing the f-number. Thus the f-number is a limiting factor in systems designed for extended-area sources or for point sources.

A practical limit for the field of view is based upon pointing accuracy: It is more difficult to acquire and or track the target with a small field of view. A theoretical limit to the smallness of the field of view is based upon the system optical blur. A point source is not imaged as a point. The optical blur is a nonzero area containing most of the flux imaged upon the focal plane. The optical blur is determined by either aberrations or diffraction. Thus, an optimum system for point sources is obtained by increasing A_c (the equivalent of reducing F) until either the optical blur matches the detector or the f-number is reduced to a practical minimum value (the theoretical minimum value is 0.5).

This can be illustrated in reference to Fig. 5.3 by rewriting Eq. (5.29) to include the parameters of detector area A_d, field of view Ω_c, and the focal length f:

$$\Omega_c = \frac{A_d}{f^2} \quad \text{and} \quad F^2 = \frac{f^2}{D^2} \qquad (5.30)$$

so that by eliminating f^2 we have

$$A_c = \frac{\pi D^2}{4} = \frac{A_d \pi}{4 F^2 \Omega_c} \qquad (5.31)$$

and then

$$\Phi_{\text{eff}} = \frac{I_s(\lambda) \, \Delta\lambda \tau_p \tau_e A_d \pi}{4 F^2 \Omega_c \beta s^2} \qquad (5.32)$$

Equation (5.32) is appropriate for point sources when the field of view Ω_c and detector area A_d are fixed by practical or theoretical considerations. The conclusions given above for the extended-area source apply equally well in this case.

EXERCISES

1. Graph the tabulated spectral response function and find the equivalent ideal spectral bandpass.

Index	Wavelength, μm	$\mathscr{R}(\lambda)$	Index	Wavelength, μm	$\mathscr{R}(\lambda)$
1	2.60	0.01	8	2.74	0.96
2	2.62	0.30	9	2.76	0.99
3	2.64	0.67	10	2.78	0.76
4	2.66	0.88	11	2.80	0.43
5	2.68	0.93	12	2.82	0.43
6	2.70	1.0	13	2.84	0.03
7	2.72	0.98	14	2.86	0.00

2. A sensor has a 2° half-angle field of view.
   (*a*) Find the incident areance [irradiance] for an extended-area source where $L_s = 1.6 \times 10^{-6}$ W cm^{-2} sr^{-1}.
   (*b*) Find the incident areance [irradiance] for a distant small-area source, where $L_s = 1.6 \times 10^{-6}$ W cm^{-2} sr^{-1}, which subtends a 1° full angle at the sensor.

3. Given the $\mathscr{R}(\lambda)$ and $L(\lambda)$ curves of Fig. E5.3:
   (*a*) Find the total sterance [radiance] L over the range $\lambda = 2.0$ to 2.3 μm.
   (*b*) Find the equivalent ideal bandwidth for $\mathscr{R}(\lambda)$.
   (*c*) Find the effective sterance [radiance] L over the range $\lambda = 2.0$ to 2.3 μm.
   Hint: Use numerical approximation techniques for integration.

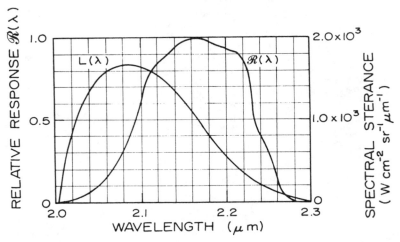

Figure E5.3

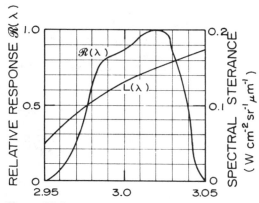

Figure E5.4

4. Given: $\mathscr{R}(\lambda)$ and $L(\lambda)$ as illustrated in Fig. E5.4:
   (a) Find the equivalent ideal bandwidth for $\mathscr{R}(\lambda)$.
   (b) Find the band sterance [radiance] L in W cm^{-2} sr^{-1} between $\lambda = 2.95$ and $\lambda = 3.05$ μm.
   (c) Find the effective band sterance [radiance] L_{eff} in units of W cm^{-2} sr^{-1}.
5. Given: A 30-cm diameter source radiating $L = 5.0 \times 10^2$ W cm^{-2} sr^{-1}. Find the sensor solid-angle field of view that is designed to maximize the flux collected at a range of 5 m.
6. Given: An $F = 2$ collector lens of 10-cm diameter. What diameter field stop (located in the focal plane) will produce a 1° half-angle field of view?
7. The flux incident upon an aperture is given by

$$\Phi = \int_\lambda \int_{\omega_s} \int_{A_s} L_s(\lambda) \cos \theta \, dA_s \, d\omega_s \, d\lambda$$

Assuming (1) the distance between A_s and A_c is large; (2) A_s and A_c are perpendicular to a line (the optical axis) connecting them; and (3) $L_s(\lambda)$ is uniform over A_s and $\Delta\lambda$, write the expression where the integrations have been carried out.
8. Given a sensor located 10 m from a source that is 2 cm in diameter. Find the projected solid angle of the sensor field of view to match the source.
9. A sensor is designed to resolve a 1-m target from a distance of 100 km. Calculate the defraction-limited aperture diameter for 2-μm wavelength response using a 0.015×0.015 mm detector. Also calculate the focal length and f-number of the optical system.

REFERENCES

1. K. Vedam, "Applications of Polarized Light in Materials Research," *Proc. Soc. Photo-Opt. Instrum. Eng.—Polarized Light*, **88**, 78–83 (1976).
2. C. L. Wyatt, *Radiometric Calibration: Theory and Methods*, Academic Press, N.Y., 1978, p. 103; see also Chap. 18.

3. F. E. Nicodemus and G. J. Zissis, "Report of BAMIRAC—Methods of Radiometric Calibration," ARPA Contract No. SD-91, Rep. No. 4613-20-R (DDC No. AD-289,375), Univ. of Michigan, Infrared Lab., Ann Arbor, MI, 1962, pp. 15–20.

4. C. L. Wyatt, *Radiometric Calibration: Theory and Methods*, Academic Press, N.Y., 1978, p. 124; also see Chap. 18.

5. F. E. Nicodemus, Ed., "Self-Study Manual on Optical Radiation Measurements. Part 1—Concepts," *Natl. Bur. Stand.* (*U.S.*), *Tech. Note* No. 910-3, 1977, p. 2.

6. P. S. Hauge, "Survey of Methods for the Complete Determination of a State of Polarization," *Proc. Soc. Photo-Opt. Instrum. Eng.—Polarized Light*, **88**, 3–10 (1976).

7. W. W. Buchman et al., "Single Wavelength Thin-Film Polarizers," *J. Opt. Soc. Am.*, **61**, 1604–1606 (1971).

8. W. L. Wolfe, Ed., *Handbook of Military Infrared Technology*, *Office of Naval Research*, Washington, DC, 1965, p. 379.

9. M. R. Holter et al., *Fundamentals of Infrared Technology*, Macmillan, N.Y., 1962, pp. 349–350.

10. E. J. Kelly et al., "The Sensitivity of Radiometric Measurements," *J. Soc. Ind. Math*, **11**, 235–257 (1963).

11. W. L. Wolfe and G. J. Zissis, Eds., *The Infrared Handbook*, Office of Naval Research, Dept. of the Navy, Washington, DC, 1978, pp. 8–29.

12. *Ibid.*, Chap. 10.

chapter 6

Detection of Radiant Flux

6.1 INTRODUCTION

A major component of the radiometric performance equation (for the signal-to-noise ratio)—the effective flux incident upon the detector based upon the source characterization, geometrical transfer, path losses, and the selective optical subsystem performance—was developed in the previous chapters. The objective of this chapter is to introduce detector figures of merit that can be used to develop the noise equivalent flux (NEF), a component of the signal-to-noise ratio equation.

In addition to the detector NEF, this chapter provides an extension of the development to include system noise equivalent sterance [radiance] and system noise equivalent areance (NEF density). These developments provide for limited (but useful) optimization based upon a criterion of minimizing the NEF using only the parameters of system throughput and detector NEF.

Generally, at the system-design level the customer is interested in sensor performance with respect to the flux as a field quantity outside or incident upon the sensor aperture. What happens inside the sensor may be of little interest. However, the design engineer needs to know how much flux is incident upon the detector in order to predict system response.

The field quantities that exist outside the sensor differ from the total flux incident upon the detector by a constant. That constant characterizes the optical subsystem parameters. Thus, the equations for minimum detectable flux are expanded to include various field quantities and system parameters.

The objective of the design of the detector subsystem is to *select* a detector capable of satisfying the system specifications of wavelength, minimum detectable flux, and frequency response.

6.2 DETECTOR PARAMETERS

The *detector* functions as a *transducer* of electromagnetic radiation in the optical bands in which radiant energy is converted into an electric signal. The choice of a detector is often the first consideration in the design of any electro-optical system.

The development of infrared detectors dates back to the pioneering discoveries of Sir William Herschel in 1800 and to the rapidly expanding area for applications to industrial and military problems. An astronomer, Sir William discovered the infrared by moving a thermometer through the sun's spectrum.[1] He found that the temperature recorded by the thermometer increased from violet to red and then continued to respond in the dark region beyond the visible. This region he called the "infra" red. For more than 100 years, little use was made of the infrared region of the spectrum, largely because of a lack of detectors of optical radiation.

The problem can be better appreciated from the fact that current state-of-the-art cryogenically cooled photoconductive detectors of the infrared are capable of detecting a power level of 10^{-15} W, which delivers signal currents of the order of 10^{-13} A from an equivalent impedance of approximately 10^{12} ohms. Infrared radiation is incoherent in nature and cannot be amplified with resonance circuits such as are commonly employed with coherent microwave signals of the same frequency (see Fig. 1.2).

A large variety of detector types are now available that respond from the ultraviolet to the far infrared. The state of the art is constantly being advanced by the improvement of existing detectors and by the introduction of new devices.

The word "sensitivity" is a term used in so many different ways that it is not very useful. The important quantity is the minimum detectable signal, which is related to the limiting noise, rather than how well a system responds to a particular input. Certain figures of merit are defined to describe detector performance under specific operating conditions; knowing them, it is possible to deduce the limits of detection. Those figures of merit that are important from a design point of view are considered here.

Four important aspects of detector performance are: (1) the noise equivalent power (NEP), (2) the absolute responsivity $\mathscr{R}$, (3) the relative spectral responsivity $\mathscr{R}(\lambda)$, and (4) the temporal response of the detector. The detector NEP is related to its ability to detect faint signals. The responsivity is an expression of output signal (usually voltage or current) per unit incident flux. Finally, the detector is inherently a band-limited device.

These four detector parameters are dependent upon the operating conditions. For example, the limiting noise mechanism, the responsivity (both absolute and spectral), and the time constant of response are dependent upon (1) the operating temperature, (2) the level of radiant background, (3) the electrical bias, (4) the electrical noise bandwidth, and (5) the physical size of the detector.

6.2.1 Detector Types

Optical system designs incorporate two basic detector types: *thermal* and *photon*. Figure 6.1 illustrates the ideal behavior of thermal and photon detectors. The distinction between them is based upon the responsive mechanism. Thermal detectors measure the rate at which *energy* is absorbed, whereas photon detectors measure the rate at which *quanta* are absorbed.

Thermal detectors are simply energy detectors, since they make use of the heating effect of radiation. Their response is dependent upon the radiant power absorbed but *independent* of the spectral content of the radiation.

Associated with all thermal detectors is some form of thermal mass that undergoes a temperature change when it absorbs radiation. Large temperature changes per unit of radiant power absorbed are associated with small thermal masses. Thus, the sensitive elements of most thermal detectors are physically small and their response rate is slow.

Examples of thermal detectors[3] are the *thermocouple*, which is based upon a thermoelectric effect; the *bolometer*. which is based upon the change in electrical resistance with temperature; and the relatively new *pyroelectric detector*,[4] which is based upon the change in polarization of a crystal when it undergoes a variation in temperature.

Photon detectors respond only to incident photons that possess more than a certain minimum energy. Thus they are *selective* detectors of optical radiation, responding only to those photons of sufficiently short wavelengths to produce charge carriers. Their response at any wavelength is proportional to the *rate* at which *photons* of that wavelength are absorbed. The number of photons per second per watt is directly proportional to the wavelength; thus, the response of a photon detector for equal amounts of radiant power per unit wavelength interval decreases as the wavelength decreases below that corresponding to the minimum energy (see Fig. 8.1).

Some types of photon detectors are: *photoemissive* detectors, which are generally used with electron multipliers and are referred to as *multiplier phototubes* (see Fig. 17.1); semiconductor *photoconductive* and *photovoltaic* detectors (see Fig. 16.2); and photographic *film*. The semiconductor detectors are composed of lead (Pb), silicon (Si), germanium (Ge), and other

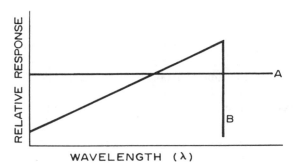

WAVELENGTH (λ)

Figure 6.1　Illustration of the ideal behavior of thermal (A) and photon (B) detectors.

substances. Various regions of wavelength response are obtained by doping these materials with certain substances. For example, germanium doped with mercury (Ge : Hg) responds to 12 μm, while germanium doped with zinc (Ge : Zn) responds to 35 μm. (Appendix D lists the elements and their chemical symbols.)

The discussion so far has described ideal detector behavior, not the actual behavior of a detector in each class. For example, most photon detectors exhibit a wavelength-dependent quantum efficiency. Thus their response is not simply a monotonically increasing function of wavelength for equal power input per unit wavelength interval but exhibits considerable deviation from this ideal. Neither do all thermal detectors exhibit a response that is completely independent of wavelength. The surface of the sensitive element of a thermal detector may have a low absorptivity in certain spectral regions, so that the energy in that region cannot be effectively utilized.

6.2.2　Detector Figures of Merit

The effective flux incident upon the detector is usually interrupted periodically with a mechanical chopper to avoid bias offset and drift problems. This results in an electric output signal that is periodic, for which ac-coupled bandpass amplifiers can be used to obtain high gain. The multiplier phototube is the only detector that has had extensive use as an unchopped detector.[6]

The detector *responsivity* and noise are measured in two separate tests and the NEP calculated from the results.[7] The rms signal voltage, V_s, is observed in response to the average incident power level; then the rms noise voltage, V_n, is observed with the detector covered so the incident power is zero.

The responsivity is defined as the ratio of the output signal voltage V_s to the radiant power Φ_i incident upon the detector, and is given by

$$\mathcal{R}(\lambda) = V_s/\Phi_i(\lambda) \qquad [\text{V rms/W rms}] \qquad (6.1)$$

Both the electric signal and the incident flux are measured in terms of the rms value of the fundamental component at the chopping frequency.[8] By rms power, in this case, is meant the average power of the waveform of the chopped radiation.[8] Equation (6.1) also holds for unmodulated (dc) flux and dc output voltage. In some cases the output is a current I_s; then the responsivity has units of A rms/W rms.

The detector *noise* must usually be measured with a preamplifier. Sometimes detectors are operated under conditions in which the detector noise is negligibly small and are therefore preamplifier noise-limited. In any case, it is important to remove the effect of amplifier gain when the noise is referred to the detector terminal. The electrical noise is measured with a relatively narrow band (tuned) voltmeter to yield the rms value at the fundamental chopping frequency.

The detector NEF is always defined in terms of radiant flux Φ_e, which has the units of watts; thus, NEP is the appropriate term. The NEP for ac operation can be defined as the ratio of the incident effective flux to the signal-to-noise ratio:

$$\text{NEP} = \frac{\Phi_{\text{eff}}}{\text{SNR}} = \frac{\Phi_{\text{eff}}}{V_s/V_n} = \frac{V_n}{\mathcal{R}} \tag{6.2}$$

where V_s/V_n is the signal-to-noise ratio and V_s/Φ_{eff} is the responsivity. The magnitude of NEP depends upon the electrical noise bandwidth of the measured noise voltage.

The meaning of NEP for a detector in ac operation is as follows: The noise equivalent power is the incident power that produces an rms output signal equal to the rms noise voltage.

The noise voltage squared (which is proportional to power) for "white" noise mechanisms (having a uniform power spectral density) is proportional to the electrical noise bandwidth. For a large class of detectors, the noise voltage squared is also proportional to the area of the detector. Thus, the rms noise voltage is proportional to the square root of both bandwidth and area.

For the special case where the noise voltage is proportional to the square root of the product of the detector area and the electrical noise bandwidth, the detectivity D^* (pronounced D-star) is a better figure of merit than NEP. D^* is defined as

$$D^*(\lambda) = (A_d \Delta f)^{1/2}/\text{NEP} \qquad [\text{cm Hz}^{1/2}/\text{W}] \tag{6.3}$$

The detectivity can be considered the inverse noise equivalent power that has been normalized for detector area and electrical noise bandwidth. Thus, the detectivity is independent of detector size and/or noise bandwidth and is representative of a particular detector type. The detectivity also has the property that numerically larger values represent superior performance, which follows the tradition that bigger is better.

The term $D^*(\lambda)$ represents the detectivity at a particular wavelength λ, while the term $D^*(T)$ is the detectivity obtained when the detector is irradiated by an unfiltered blackbody source, a convenient measurement technique. In this case, the incident power used in Eqs. (6.1) through (6.3) is obtained as the integral of Planck's equation for all wavelengths (see Chap. 9). Correction factors are used to convert $D^*(T)$ to $D^*(\lambda)$.

D^* can also be defined in terms of the responsivity by combining Eqs. (6.2) and (6.3)

$$D^*(\lambda) = \frac{\mathscr{R}(\lambda)(A_d \Delta f)^{1/2}}{V_n} \qquad \left[\text{cm Hz}^{1/2}/\text{W} \right] \qquad (6.4)$$

6.3 SYSTEM PARAMETERS

The calculation of system figures of merit is based upon detector figures of merit and optical system constants. The detector NEP can be expressed in terms of detector figures of merit of responsivity and/or D^*, noise voltage and electrical noise bandwidth as follows:

$$\text{NEP}(\lambda, \text{det}) = V_n / \mathscr{R}(\lambda) \qquad [\text{W}] \qquad (6.5)$$

or

$$\text{NEP}(\lambda, \text{det}) = \frac{(A_d \Delta f)^{1/2}}{D^*(\lambda)} \qquad (6.6)$$

6.3.1 Noise Equivalent Power

Equations (6.5) and (6.6) must be modified to yield the NEP as a field quantity at the sensor entrance aperture. This is necessary in order to take into account the optical losses and noise effects associated with dc restoration as follows:

$$\text{NEP}(\lambda, \text{sys}) = \frac{\beta V_n \sqrt{2}}{\mathscr{R}(\lambda) \tau_e} \qquad (6.7)$$

or

$$\text{NEP}(\lambda, \text{sys}) = \frac{\beta (2 A_d f_2)^{1/2}}{D^*(\lambda) \tau_e} \qquad (6.8)$$

where β is the chopping factor and τ_e is the optical efficiency at wavelength λ. The factor $\sqrt{2}$ in Eqs. (6.7) and (6.8) results from an increase in the noise because of the effect of coherent rectification upon the electrical noise bandwidth. The electrical noise bandwidth Δf is replaced in Eq. (6.8) by the low-pass filter cutoff frequency f_2 that is used with coherent rectification. Coherent rectification is utilized to obtain dc restoration in chopped systems resulting in a mean dc output that is proportional to the mean incident flux. This is discussed in detail in Sec. 7.5.

Equations (6.7) and (6.8) yield NEP at the indicated wavelengths but imply nothing about the detectable power at any other wavelength.

Example 1: Find the system NEP for a band radiometer that utilizes the coherent rectification scheme for dc restoration.

Given: The detector area is 7.85×10^{-3} cm^2 (1 mm diameter), $D^*(2.7$ μm$) = 9 \times 10^9$ cm Hz$^{1/2}$/W, optical efficiency is $\tau_e = 0.62$ at 2.7 μm, chopping factor $\beta = 3$, low-pass $f_2 = 10$ Hz.

Basic equation:

$$\text{NEP}(\lambda, \text{sys}) = \beta(2A_d f_2)^{1/2}/D^*(\lambda)\tau_e \quad [\text{W}] \quad (6.8)$$

Assumptions: (1) $D^*(\lambda)$ is constant over the spectral bandwidth. (2) The limiting noise is detector, not preamplifier, noise. (3) The flux incident upon the entrance aperture originates within the sensor field of view and from within the sensor spectral band.

Solution: The NEP at 2.7 μm is

$$\text{NEP}(2.7\,\mu\text{m}) = \frac{3\{2 \times 7.85 \times 10^{-3}[\text{cm}^2] \times 10[\text{Hz}]\}^{1/2}}{9 \times 10^9[\text{cm Hz}^{1/2}] \times 0.62}$$

$$= 2.13 \times 10^{-10} \text{ W}$$

Note: The system NEP is larger than the detector NEP because of system losses. ∎

6.3.2 Noise Equivalent Sterance [Radiance]

The flux incident upon the sensor aperture may be expressed as a field quantity using the most general radiometric entity, the sterance [radiance], which has the units of W cm^{-2} sr^{-1}. The sterance [radiance] may be thought of as the ratio of the power in a beam to the optical throughput Υ of the beam [see Eq. (3.21)].

The throughput of the sensor entrance aperture can be considered as a beam of radiant energy, and the flux can be viewed as a field quantity anywhere in the beam. The sensor throughput is also given by the invariance theorem, in terms of the focal-plane f-number F and the detector field stop area A_d as

$$\Upsilon = A_c\Omega_c = A_d\pi/4F^2 \quad [\text{cm}^2 \text{ sr}] \quad (6.9)$$

where A_c is the effective collecting aperture area, Ω_c is the sensor projected field of view, and A_d is the detector area that serves as the field stop.

The noise equivalent sterance [radiance], NES [NER], is obtained from Eqs. (6.7) or (6.8) and (6.9) as

$$NES(\lambda) = \frac{NEP(\lambda, sys)}{A_c \Omega_c} = \frac{4F^2 NEP(\lambda, sys)}{\pi A_d} \qquad [\text{W cm}^{-2}\,\text{sr}^{-1}] \quad (6.10)$$

which in terms of detector responsivity $\mathscr{R}$ and noise voltage V_n [Eq. (6.7)], is

$$NES(\lambda) = \frac{\sqrt{2}\left(4\beta F^2 V_n\right)}{\pi \mathscr{R}(\lambda)\tau_e A_d} \qquad (6.11)$$

or in terms of D^* [Eq. (6.8)] as

$$NES(\lambda) = \frac{4\beta F^2 \left(2f_2/A_d\right)^{1/2}}{\pi D^*(\lambda)\tau_e} \qquad (6.12)$$

Equations (6.11) and (6.12) can be used for sensitivity analysis to optimize the noise equivalent sterance [radiance] of a system based upon the throughput parameters of the optical system and the detector parameters. These equations are of limited value because they do not contain terms to describe the source, media losses, or selective properties of the optical subsystem. Consequently, minimizing the NES does not necessarily maximize the signal-to-noise ratio. The NES [NER], Eq. (6.11), is illustrated with an example that shows it is independent of the field of view.

Example 2:   Find the noise equivalent sterance [radiance] for a band radiometer that utilizes coherent rectification for dc restoration.

Given: Detector current responsivity is $\mathscr{R}_c = 2$ A/W at 5 μm. The detector and associated electronics are cryogenically cooled to 78 K, greatly reducing the background; thus, the limiting noise is thermal (rather than photon), originating from the load resistance of 1×10^9 ohms (see Fig. 18.2). The thermal noise voltage is given by the Johnson-Nyquist equation, $V_n = (4kTR_d)^{1/2}$ (see Sec. 7.5.1). The system is chopped ($\beta = 3$), the optical efficiency is $\tau_e = 0.3$, the optical system f-number is $F = 2$, the noise bandwidth $f_2 = 62$ Hz, and the detector area $A_d = 7.85 \times 10^{-3}$ cm^2.

Basic equations:

$$V_n = (4kTR_d)^{1/2} \qquad [\text{V/Hz}^{1/2}]$$

$$NES(\lambda) = \frac{\sqrt{2}\left(4\beta F^2 V_n\right)}{\pi \mathscr{R}(\lambda)\tau_e A_d} \qquad [\text{W cm}^{-2}\,\text{sr}^{-1}] \quad (6.11)$$

Assumptions: (1) The responsivity is constant throughout the radiometer spectral band. (2) The flux incident upon the aperture originates from within the sensor field of view and from within the spectral band.

Solution: The thermal noise voltage is

$$V_n = (4kTR_df_2)^{1/2}$$
$$= \{4 \times 1.38 \times 10^{-23}[\text{J K}^{-1}] \times 78[\text{K}]$$
$$\times 1 \times 10^9[\text{ohms}] \times 62[\text{Hz}]\}^{1/2}$$
$$= 1.63 \times 10^{-5} \text{ V rms}$$

for a low-pass cutoff frequency of 62 Hz. The noise equivalent sterance [radiance] is obtained by

$$\sqrt{2}(4\beta F^2V_n) = \sqrt{2} \times 4 \times 3 \times 2^2 \times 1.63 \times 10^{-5} \text{ V} = 1.11 \times 10^{-3} \text{ V}$$
$$\pi\mathscr{R}(\lambda)\tau_e A_d = \pi \times 2.0[\text{A/W}] \times 1 \times 10^9[\text{ohms}]$$
$$\times 0.3 \times 7.85 \times 10^{-3}[\text{cm}^2]$$
$$= 1.48 \times 10^7 \text{ V cm}^2 \text{ sr/W}$$
$$\text{NES}(5.0 \ \mu\text{m}) = \frac{1.11 \times 10^{-3}}{1.48 \times 10^7} = 7.47 \times 10^{-11} \text{ W cm}^{-2} \text{ sr}^{-1}$$

Note: The magnitude of the field of view can be traded off for collector area without changing NES provided the throughput is maintained constant. ∎

6.3.3 Noise Equivalent Areance [Irradiance]

As indicated in Sec. 4.2.1, the sterance [radiance] is the most appropriate entity with which to describe an extended-area source. On the other hand, according to Sec. 4.2.2, a small-area source (point source) at an unknown distance is best described in terms of the incident flux density, the areance [irradiance] E, which has the units of W/cm^2. The areance [irradiance] is related to the sterance [radiance] L for a sensor by

$$E = L\Omega_c \quad [\text{W/cm}^2] \tag{6.13}$$

where Ω_c is the sensor solid-angle field of view.

The noise equivalent areance [irradiance] is commonly referred to by the acronym NEFD, taken from the words "noise equivalent flux density." The NEFD for a distant small-area source is obtained using Eq. (6.7) as

$$\text{NEFD}(\lambda) = \frac{\sqrt{2} \ \beta V_n}{\mathscr{R}(\lambda)\tau_e A_c} \tag{6.14}$$

and from Eq. (6.8) as

$$\text{NEFD}(\lambda) = \frac{\beta(2A_d f_2)^{1/2}}{D^*(\lambda)\tau_e A_c} \tag{6.15}$$

where A_c is the area of the collecting aperture.

Equations (6.14) and (6.15) are not very useful for optimization and sensitivity analysis for the reasons given in Sec. (5.10). Combining Eqs. (5.33) and (6.14), we obtain

$$\text{NEFD}(\lambda) = \frac{\sqrt{2}\,\beta V_n 4 F^2 \Omega_c}{\mathcal{R}(\lambda)\tau_e A_d \pi} \tag{6.16}$$

in terms of the detector responsivity and noise voltage, where Ω_c is the field-of-view solid angle and A_d is the detector area that serves as a field stop.

Combining Eqs. (5.33) and (6.15) we obtain

$$\text{NEFD}(\lambda) = \frac{\beta(2A_d f_2)^{1/2}}{D^*(\lambda)\tau_e A_c} = \frac{\beta(2A_d f_2)^{1/2} 4 F^2 \Omega_c}{D^*(\lambda)\tau_e A_d \pi}$$

$$= \frac{4 F^2 \beta \Omega_c}{D^*(\lambda)\tau_e \pi}\left(\frac{2 f_2}{A_d}\right)^{1/2} \tag{6.17}$$

in terms of D^*, where Ω_c is the field-of-view solid angle and A_d is the detector area that serves as the field stop. In general, the conclusions given above for NES apply here when there is a fixed field-of-view requirement for a point-source system.

6.3.4   Noise Equivalent Spectral Flux

Each of the entities described in Secs. 6.3.1 through 6.3.3 may be applied to any sensor that makes a measurement of the *total* radiation over a specified band $\Delta\lambda$ such as a radiometer. However, a spectrometer measures the power spectral density function (or simply the "spectrum"), which has the units of flux per unit wavelength (or wavenumber).

The analysis for spectral NEF is essentially as given above for band flux. In the case of a spectrometer, it is the *instantaneous spectral bandwidth* that is used to calculate the total flux on the detector.

EXERCISES

1. A sensor system responds to 1×10^{-9} W/cm^2 incident flux with 1.57 V output. The dark noise is 1.27 mV. Find the system noise equivalent flux density (NEFD).
2. For the sensor of Exercise 1, find the system NEP, given the collecting aperture diameter is 10 cm.

3. A 1×1 mm detector is tested with a preamplifier that has a gain of 1000. The radiant source is modulated with a light chopper at 150 Hz. Assume a chopping factor of 3. Two tests are conducted, and the results are as follows: (i) The output voltage is 2.3 V rms when the incident areance [irradiance] is 1.6×10^{-8} W/cm² (peak power density). (ii) The detector is shielded from any modulated energy, and the noise voltage is measured using a tuned voltmeter. The voltmeter has a noise bandwidth of 10 Hz, and the output noise voltage is found to be 1.2 mV rms.

Assuming the preamplifier contributes zero noise, find (a) the detector responsivity, and (b) the detector D^*.

4. Find the NEP(λ, det) for a 0.5×0.5 cm detector, given that $D^*(\lambda) = 1 \times 10^{10}$ cm Hz$^{1/2}$/W and the noise bandwidth is 100 Hz.

5. Given: $D^*(\lambda = 5.0 \ \mu\text{m}) = 1 \times 10^{13}$ cm Hz$^{1/2}$/W for a detector of area $= 7.85 \times 10^{-3}$ cm². Find: NEP($\lambda = 5.0 \ \mu$m) for an electrical noise bandwidth of 1000 Hz.

6. Given: Detector NEP(λ, det) $= 1 \times 10^{-13}$ W, chopping factor $\beta = 3$, optical system efficiency $\tau_e = 0.5$, throughput $\Upsilon = A\Omega = 10^{-3}$ cm² sr. Find the system NES for an extended area source.

7. A system has a detector with responsivity $\mathscr{R} = 10^6$ V/W, noise voltage $V_n = 10$ μV, chopping factor $\beta = 3$, optical efficiency $\tau_e = 0.5$, and collector diameter 25 cm and uses coherent rectification. Find the system NEFD.

8. A detector of area 0.25 cm² is irradiated by $E = 1 \times 10^{-6}$ W/cm² rms. The resultant signal voltage is 1.5 mV rms. The detector noise voltage is measured at 1×10^{-5} V rms in a 10-Hz bandwidth. Find the detector responsivity $\mathscr{R}$, NEP(λ, det), and D^*.

REFERENCES

1. A. R. Laufer, "Preface," *Proc. IRE*, **47** (1959), special issue on infrared physics and technology, p. 1415.
2. R. A. Smith et al., *The Detection and Measurement of Infra-red Radiation*, Oxford University Press, London, 1957, pp. 7–16.
3. *Ibid.*, p. 56–118.
4. R. J. Phelan, Jr., and A. R. Cook, "Electrically Calibrated Pyroelectric Optical-Radiation Detector," *Applied Optics*, **12**, 2494–2500 (1973).
5. R. A. Smith et al., *The Detection and Measurement of Infra-red Radiation*, Oxford University Press, London, 1957, pp. 119–171.
6. E. H. Eberhardt, "Noise in Photomultiplier Tubes," *IEEE Trans. Nuclear Sci.*, **Ns-14**, 7–14 (1967).
7. R. C. Jones et al., "Standard Procedure for Testing Infrared Detectors and Describing Their Performance," Office of Director of Defense Research and Engineering, Washington, DC, 1960, pp. 10–13.
8. R. J. Smith, *Electronics: Circuits and Devices*, Wiley, N.Y., 1973, p. 83.

chapter 7

Signal-Conditioning Electronics

7.1 INTRODUCTION

The objective of this chapter is to consider signal-processing techniques and noise mechanisms associated with the signal-conditioning subsystem. This chapter emphasizes electrical parameters, which are described in general terms that relate to overall system performance. The electrical subsystem normally determines the temporal characteristics of the system. Optional signal-conditioning techniques involve trade-offs to minimize the limiting effects of noise.

The objective of the design of the signal-conditioning subsystem is to establish the noise/information bandwidth, a term required to solve the SNR radiometric performance equation. There are at least two classifications of application goals: (1) for radiometric systems, to provide the best possible estimate of the source flux for stationary processes and (2) to establish the sample rate for nonstationary sources encoded with information.

The best that any radiometric or spectrometric instrumentation system can accomplish is to provide an *estimate* of the magnitude of the source flux.[1] The electrical subsystem is designed to provide optimum signal processing to improve that estimate so far as the limiting effects of noise are concerned.

Optical communication links and remote-sensing systems provide a wide variety of functions such as target detection,[2] scene representation (radar and television), and fiber optics data links. Such systems are optimized through the use of matched filters. In the statistical theory of filtering, an *optimum filter* is the filter that operates with the minimum mean square error.[3]

The electronics must also provide enough gain to yield useful output levels and operate over a dynamic range sufficient to provide linear response over the expected flux levels. The signal-conditioning electronics also provide for electrical interface with the radiant detector and in some cases are the source of the limiting noise.

A major goal of the feasibility study is to determine the information bandwidth and required sampling rates and to determine the limiting noise.

7.2 INFORMATION BANDWIDTH

The *integration time* T_d is defined as the amount of time available to obtain a measurement in which there is little or no change in the radiant levels. It can be visualized by assuming the signal to be an ideal square-wave pulse of width τ as illustrated in Fig. 7.1. In this case the pulse duration τ is the integration time T_d. The objective of the measurement is to estimate the magnitude of the pulse. The sensor response to this square wave is given by

$$f(t) = 100[1 - \exp(-T_d/T_c)] \qquad [\%] \qquad (7.1)$$

where T_c is the system time constant.

The solution to Eq. (7.1) is obtained by expressing the integration time in terms of the system time constant. For example, when

$$T_d = 5T_c \qquad [s] \qquad (7.2)$$

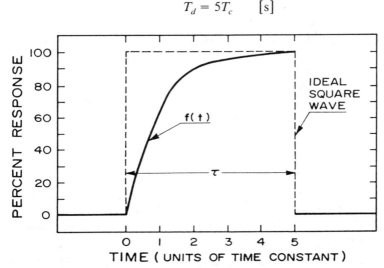

Figure 7.1 Time response to a square-wave signal of a low-pass *RC* filter where $\tau = 5T_c$.

the system output achieves 99.3% of the source amplitude as shown in Fig. 7.1, which represents a nominal design value for systems that require *accuracy* in measuring the pulse magnitude in analog systems.

The electrical frequency response is related to the time constant for simple *RC* networks by

$$f_2 = \frac{1}{2\pi T_c} \quad [\text{Hz}] \tag{7.3}$$

where f_2 is the upper frequency limit for a low-pass filter.

7.2.1 Sampling Rate

Modulated radiant sources are generally continuous in time but are not necessarily periodic. The electro-optical sensor provides at best a periodic sample of the source flux. A complete description of the temporal properties of a modulated radiant signal does not require that the amplitude be known at every instant of time. This is a consequence of the fact that all signals are bandwidth-limited[4] and is the basis of the sampling theorem that is demonstrated by the following argument.

The lowest frequency possible in a sample set of duration T_s s is $1/T_s$ Hz. Suppose the band-limited signal is known to contain no energy at frequencies above f_m. The signal can be examined in terms of a Fourier series expansion, and the maximum number of harmonics present is $f_m T_s$. In order to completely describe the signal, we need only find the sine and cosine coefficients, or the magnitude and phase angle, for each term. Thus a total of $2 f_m T_s$ coefficients are required, and they can be evaluated from the same number of samples.

The effect of sampling periodically, t_s s apart, where the sample rate $f_s = 1/t_s$ is visualized by considering that the signal modulates the amplitude of a periodic pulse train as shown in Fig. 7.2. The process of amplitude modulation can be represented as a multiplication process in which the product of the original signal and the sample function is obtained.

A simple switch periodically switched off or on, or an optical chopper, can be expressed by a Fourier series expansion as

$$\tfrac{1}{2}\left[1 + \sum_n b_n \sin(n\omega_s t)\right] \tag{7.4}$$

where b_n is the amplitude and ω_s is the sampling frequency. The original optical signal contains a dc component, as all optical signals must, since negative flux is impossible. The information bandwidth can be represented by a sine-wave modulation at the maximum frequency f_m. This complex signal can be expressed as

$$a + b \sin \omega_m t \tag{7.5}$$

where a is the amplitude of the dc component and b is the peak amplitude

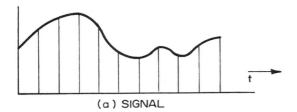

(a) SIGNAL

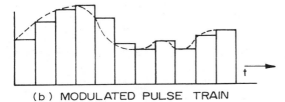

(b) MODULATED PULSE TRAIN

Figure 7.2 Illustration of modulated signal (a) and a modulated sample pulse train (b).

of the sine-wave component at the frequency $\omega_m = 2\pi f_m$. The sampling process yields the *product* of Eqs. (7.4) and (7.5):

$$\frac{1}{2}a + \frac{b}{2}\sin(\omega_m t) + \frac{a}{2}\sum_n b_n\sin(n\omega_s t) + \frac{b}{2}\sin(\omega_m t)\sum_n b_n\sin(n\omega_s) \quad (7.6)$$

which is

$$\frac{1}{2}a + \frac{b}{2}\sin(\omega_m t) + \frac{a}{2}\sum_n b_n\sin(n\omega_s t)$$

$$+ \frac{b}{4}\sum_n b_n\left[\cos(\omega_m - n\omega_s)t - \cos(\omega_m + n\omega_s)t\right] \quad (7.7)$$

where the identity

$$\sin x \sin y = \tfrac{1}{2}[\cos(x - y) - \cos(x + y)] \quad (7.8)$$

is used and where n ranges from unity to $f_m T_s$.

The first two terms of Eq. (7.7) represent one-half the dc and modulation terms. The third term represents $n = f_m T_s$ terms of the sample frequency f_s and its odd harmonics, $3f_s$, $5f_s$, etc. The fourth term represents the sidebands at $f_s \pm f_m$, $3f_s \pm f_m$, $5f_s \pm f_m$, etc. This is depicted in Fig. 7.3, where it is seen that the information can be retrieved using an ideal filter provided that $f_m < f_s - f_m$ so the sidebands do not overlap. Thus, according to the sampling theorem,[4] the sampling frequency must be at least

$$f_s \geq 2f_m \quad [\text{Hz}] \quad (7.9)$$

It is necessary to set the sample rate at 3 to 5 times the maximum frequency because of the nonideal characteristics of practical filters.

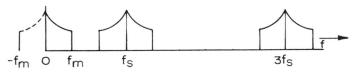

$-f_m$ 0 f_m f_s $3f_s$

Figure 7.3 Illustration of sampling of a continuous signal and resulting sidebands. (Note: The even terms are missing for square-wave sampling.)

7.2.2   The Spectrometer as a Sample System

The sample rate for any spectrometer is the scan rate S, since each spectral component is sampled only once each scan. The radiant energy incident upon the detector contains modulation components inherent in the source and those resulting from the action of the scanner.

A sequential spectrometer is one that samples each incremental wavelength one after the other; thus, the optical signal is modulated at a frequency of NS Hz, where N is the number of incremental elements resolved for each scan. The frequency spectrum of the output signal is the product of the scan frequency and the source modulation frequencies in accordance with Eq. (7.7). Thus, the maximum frequency in the detector output is

$$\text{Maximum frequency} = NS + f_m \quad [\text{Hz}] \tag{7.10}$$

where f_m is the maximum modulation frequency in the flux beam.

The use of a light chopper in the sequential spectrometer constitutes a second sampling function for which the sample rate corresponds to the chopping frequency. According to the sampling theorem, the chopper frequency must be two times the highest modulation component, which for a sequential spectrometer is

$$f_c \geq 2(NS + f_m) \quad [\text{Hz}] \tag{7.11}$$

Example 1:   A sequential spectrometer is to be used to measure $N = 50$ resolution elements. Determine the scan rate, chopper frequency, dwell time, time constant, and frequency response.

Given: The highest radiant source modulation frequency is 2 Hz.

Basic equations:

$$T_d = 5T_c \quad [\text{s}] \quad \text{(For magnitude accuracy)} \tag{7.2}$$

$$f_2 = \frac{1}{2\pi T_c} \quad [\text{Hz}] \quad \text{(Frequency response)} \tag{7.3}$$

$$f_s \geq 2f_m \quad [\text{Hz}] \quad \text{(Sampling rate)} \tag{7.9}$$

$$f_c \geq 2(NS + f_m) \quad [\text{Hz}] \quad \text{(Chopper frequency)} \tag{7.11}$$

Assumptions: The limiting-case solution is obtained by setting the sampling frequency at $2f_m$.

Solution: According to the sampling theorem, the scan rate S, which for a sequential spectrometer is the sample rate f_s, must be at least two times the highest modulation frequency, or

$$S = f_s \geq 2f_m \geq 2 \times 2 \geq 4 \text{ Hz}$$

The integration or dwell time, for equal resolution increments, is

$$T_d \leq \frac{1}{NS} \leq \frac{1}{50 \times 4} \leq 0.005 \text{ s}$$

which is the duration time for each resolution increment.
The system time constant for accurate magnitude is calculated by Eq. (7.2):

$$T_c = T_d/5 = 0.005/5 = 0.001 \text{ s}$$

The corresponding information bandwidth is

$$f_2 = 1/(2\pi T_c) = 1/(2\pi \times 0.001) = 159 \text{ Hz}$$

The scanner produces modulation components at $SN + f_m = 202$ Hz, so the chopper frequency must be

$$f_c \geq 2(SN + f_m) \geq 404 \text{ Hz}$$

The bandpass is 404 ± 159 Hz. ∎

A radiometer can be considered a degenerate case of the sequential spectrometer for which $N = 1$. In this case the integration (or dwell) time is simply the inverse of the sample rate f_s:

$$T_d = \frac{1}{NS} = \frac{1}{f_s} \qquad [\text{s}] \qquad (7.12)$$

which must be established in accordance with the sampling theorem as given above, that is, $f_s \geq 2f_m$.

For the case of a Michelson interferometer spectrometer, each monochromatic element of the radiant source is transformed into an electrical frequency according to the relation[5]

$$f = v\bar{\nu} \qquad [\text{Hz}] \qquad (7.13)$$

where v is the rate of change of path difference (cm/s) and $\bar{\nu}$ is the wavenumber (reciprocal centimeters, cm^{-1}) so that the time constant depends upon the maximum frequency obtained for the maximum wavenumber, $\bar{\nu}_m$, for a given rate.

7.2.3 Aliasing

Rather serious errors occur when the continuous signal contains some unsuspected components at frequencies higher than half the sampling rate. This phenomenon is known as *aliasing*.[6] The effect of aliasing is to

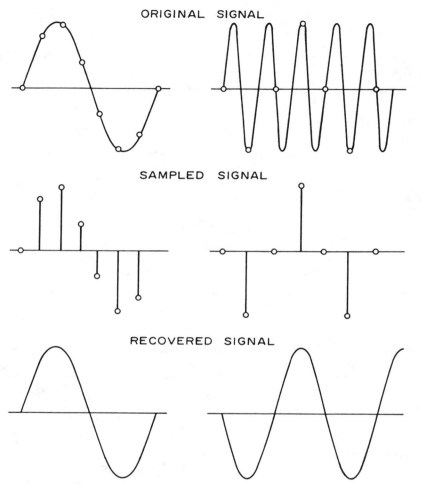

Figure 7.4 Illustration of aliasing error. Left side, the sampling frequency is 7 times the original frequency; right side, the sampling frequency is 7 / 5 of the original frequency.

introduce false frequencies into the output that were not present in the original signals.

Aliasing is illustrated in Fig. 7.4. The sample frequency at the left in Fig. 7.4 is chosen to be about 7 times that of the original sine wave. The recovered version looks very similar to the original.

The sample frequency at the right in Fig. 7.4 is chosen to be about 7/5 of the original frequency, a clear violation of the sampling theorem. In this case the recovered version does not resemble the original continuous waveform. In fact, the false signal frequency of the recovered signal is exactly equal to the difference between the original continuous signal and the sample frequency.

All signals are band-limited by the physical phenomena giving rise to them. However, real processes do not exhibit ideal band-limited spectra. The signal must be band-limited before sampling takes place to avoid aliasing. Spatial filtering of electro-optical signals results from the response characteristics of the optical system (see Chap. 12) and provides for anti-aliasing band-limiting.

7.2.4 Electrical Noise Bandwidth

The limiting noise in electro-optical systems can arise from a number of different mechanisms and depends upon the detector and preamplifier characteristics and the information bandwidth. There are two idealized noise spectral density functions: white noise and "colored" noise.

White noise exhibits a *uniform* noise spectrum for which the mean square noise voltage is given by

$$\overline{V^2} = c \int_{f_1}^{f_2} df = c(f_2 - f_1) = c \Delta f \tag{7.14}$$

where the constant c is independent of frequency and depends upon the noise mechanism. The rms noise in the output of a uniform noise source is therefore proportional to the square root of the bandwidth.

Colored noise exhibits a *nonuniform* spectrum, and there are two important cases:

Case 1 The rms noise of a system that has a $1/f$ noise distribution tends to be independent of system bandwidth. This occurs in high-gain dc-coupled systems. Drift in the input stages may be viewed as a very low frequency noise.

Generally, light choppers are used with systems that are subject to detector bias or preamplifier offset drift. This results in periodic flux and ac output signal and permits the use of amplifiers that block the dc and associated drift. The chopper samples the incoming flux beam in accordance with Eq. (7.7), transforming the frequency spectrum of the signal to the sample frequency as illustrated in Fig. 7.5, where the radiant energy is depicted as having a positive mean value modulated at a frequency f_m. The chopped detector output spectrum results in a carrier at f_c proportional to the mean flux value, and sidebands at $f_c \pm f_m$, each proportional to half the

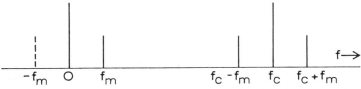

Figure 7.5 Illustration of the transformation of the signal spectrum through the use of a light chopper.

amplitude of f_m, in accordance with Eq. (7.7). The chopping frequency f_c is chosen to provide the most favorable detector-preamplifier noise characteristic. The great majority of systems fall into this category. The use of light choppers effectively eliminates the $1/f$ noise. The resultant noise is characteristic of frequencies in the neighborhood of the carrier frequency.

Case 2   A special case of colored noise occurs when the mean-square noise is proportional to the square of the frequency. This occurs in systems that are preamplifier noise limited and for which the input capacitance causes the preamplifier loop gain to roll up at 6 dB/octave (see Sec. 18.6.2). In this case the mean-square noise is given by

$$\overline{V_n^2} = c \int_{f_1}^{f_2} f^2 \, df \tag{7.15}$$

The carrier frequency f can be considered constant over the bandwidth when the bandwidth is relatively narrow. Then

$$\overline{V_n^2} \simeq c \int_{f_1}^{f_2} f^2 \, df = cf^2(f_2 - f_1) = cf^2 \Delta f \tag{7.16}$$

and the mean-square noise is therefore proportional to the bandwidth Δf and to the frequency squared.

For wide-band systems (where f cannot be considered constant) the mean-square noise is given by

$$\overline{V_n^2} = c \int_{f_1}^{f_2} f^2 \, df = c \frac{f_2^3 - f_1^3}{3} \tag{7.17}$$

for which the rms noise voltage is proportional to

$$V_n \approx f_2^{3/2} \quad [\text{V rms}] \tag{7.18}$$

where $f_2 > f_1$.

Each of the noise mechanisms associated with the detector-preamplifier subsystem must be evaluated in order to determine the optimum chopping frequency and the resulting system noise. This probably constitutes the most difficult aspect of a feasibility study. The outcome of such a study depends upon the detector type, the operating temperature and background, and the preamplifier noise. This is covered in detail in Chap. 18.

7.3 DATA RATES

Pulse code modulation (PCM) is a time-sampled or multiplex data system that is appropriate for use with optical communication systems. As in any sampled system, the information bandwidth requirements dictate the sample rates in accordance with the sampling theorem. However, the communication link bandwidth is based upon the criterion that only the presence or absence of a pulse is required rather than that pulse magnitude accuracy is necessary. One level represents a 1 bit; a second level represents a 0 bit. The actual voltage level that represents the 1 and the 0 may vary with the particular design; however, the transmitter, transmission path, detector, and detection equipment need handle only two levels. Thus, linearity of these components is not important. In the absence of noise, the accuracy of the data received is limited only by the end instruments: the multiplexer, encoder, and decoders. Essentially error-free transmission of PCM is possible with a signal-to-noise ratio of 16 to 18 dB.

Each sample is taken from the analog signal channel and "encoded" into a series of N binary digits, referred to as a "word," which describes the analog level of the sample. The output of the encoder, which is more commonly referred to as an analog-to-digital (A/D) converter, is a voltage waveform that represents the 1 and 0 bits.

The accuracy of the data is determined by the bits per word. An N-bit system refers to a system capable of handling N information bits per word. The number of discrete values, M, obtained for an N-bit system is given by

$$M = 2^N \qquad (7.19)$$

Using 12 bits, for example, 4096 discrete numbers, 0 through 4095, can be transmitted. The accuracy of such a system is given by

$$\text{Accuracy} = \left(1 - \frac{1}{M}\right) \times 100\% \qquad (7.20)$$

which for a 12-bit system is 99.976%.

Encoders are capable of 16-bit accuracy. Another advantage is the capability of handling a large number of data channels over a single link with little increase in hardware over that required for several channels. A series of n words and synchronizing bits comprising the sampled and multiplexed data for n data channels is referred to as a "frame."

The data-stream format must include extra pulses for synchronization. Synchronization patterns generally comprise 2 to 5% of the data transmitted.[7] Word synchronization patterns typically use one or more bits between words. Where more than one data channel is being encoded, the frame synchronization uses one word per frame.

In any time-division multiplex system, the first consideration is the sampling rate required for each channel of information to be transmitted.

As indicated above, the link frequency response does not have to faithfully reproduce the pulse shape or voltage levels; it is only necessary to determine the presence or absence of a pulse. It has been empirically determined that the link bandwidth should be from one-half to three-fourths of the bit rate frequency.[8]

The following example illustrates the application of PCM for a multichannel fiber optics link.

Example 2:   Find the (PCM) link frequency response required.

Given: Ten data channels are to be multiplexed, the information bandwidth for each channel is 5 kHz, and the accuracy required is 99%.

Basic equations:

$$M = 2^N \qquad\qquad (7.19)$$
$$\text{Accuracy} = (1 - 1/M) \times 100\% \qquad\qquad (7.20)$$

Assumptions: None

Solution: The number of bits required for 99% accuracy is given by solving Eq. (7.20) for M:

$$M = \frac{1}{1 - \text{accuracy}/100} = 100$$

and by solving Eq. (7.19) for the number of bits N:

$$N = \text{int}\left(\frac{\log M}{\log 2} \right) + 1 = 7 \text{ bits}$$

This is accomplished by taking the logarithm of both sides of Eq. (17.19) and rounding off to the next higher value. In this case it is appropriate to use a standard 8-bit system, which will provide an accuracy better than required. As a check, Eqs. (7.19) and (7.20) are solved again for $N = 8$ to obtain

$$M = 2^8 = 256$$

and

$$\text{Accuracy} = (1 - 1/256) \times 100 = 99.61\%$$

The sample rate corresponds to the frame rate, since each channel is sampled once per frame. According to the sampling theory, two samples per cycle of the maximum data channel frequency are required; this results in a frame rate of $2 \times 5 = 10$ kHz.

Two bits are used for word synchronization, so the word length is 10 bits. One word is used for frame synchronization, so each frame contains 11 words. Thus, the bit rate is given by

$$10{,}000 \, \frac{\text{frames}}{\text{s}} \times 10 \, \frac{\text{bits}}{\text{word}} \times 11 \, \frac{\text{words}}{\text{frame}} = 1.1 \times 10^6 \text{ bits/s}$$

Finally, the link frequency response is given by

$$F_2 = 0.5 \times \text{bit rate} = 550 \text{ kHz} \qquad \blacksquare$$

Data restoration is a process by which the original signal is derived from the sample function. This is accomplished using decommutation and digital-to-analog (D/A) converters with appropriate low-pass filters. Provided no frequencies exist in the channel data at rates higher than the Nyquist rate (half the sample rate), the output data will perfectly reproduce the original analog data.[9]

7.4 DYNAMIC RANGE

The dynamic range of a sensor relates to the range of input flux over which a useful output is possible. For linear systems the dynamic range is given by the ratio of the full-scale system output voltage to the voltage change produced by the NEF. For nonlinear systems the flux dynamic range is obtained by substituting the limiting voltages into the nonlinear transfer function.

The ultimate dynamic range that a system is capable of exhibiting is determined as the ratio of the flux levels corresponding to the ratio of the detector-preamplifier full-scale output voltage to the preamplifier rms output noise voltage.

Example 3: Find the dynamic range of a radiometer.

Given: The radiometer has four channels to increase the dynamic range with full-scale output to 10 V each. The relative channel gain is the factor 10. The output noise of the high-gain-channel is measured at 56 mV rms.

Assumptions: The system is assumed to be linear, so the dynamic range of the output corresponds to that of the input flux.

Solution: The relative gain of the high- to low-gain channels is 10^3. Full-scale output on the low-gain channel (10 V) is therefore equivalent to $10 \times 10^3 = 10^4$ V on the high-gain channel; thus, the dynamic range is given by

$$10^4/(5.6 \times 10^{-3}) = 1.79 \times 10^6 \qquad \blacksquare$$

7.5 NOISE

The ultimate limit to the detection of faint levels of radiant flux is set by random fluctuations of a fundamental nature. These fluctuations are termed "noise." The noise appearing at the output of an electro-optical system may arise from any number of sources that can be classified in one of three categories: (1) noise in the radiating background, (2) noise in detectors, and (3) noise from circuits.

The phenomenon of background noise results from the random arrival of photons. It is of special interest because it is the factor that ultimately limits the performance of detectors. A detector that is limited by random noise originating in the radiation background is referred to as a BLIP, for background-limited infrared photoconductor. The incident background flux is proportional to detector area, and the resultant noise is proportional to the square root of the detector area; hence, the detectivity D^* is a preferred figure of merit for detectors operated under BLIP conditions.

7.5.1 Noise in Resistive Elements

Fluctuations in the concentration and motion of the current carriers in a resistive conductor or semiconductor material give rise to fluctuations in the output signal of the system. This includes detectors and signal-conditioning circuit components. The system designer must exercise careful design and component selection to reduce the noise generated in the circuits to a level below that of the background noise where possible.

Many detectors of optical radiation and other circuit components exhibit *thermal noise*, a form of noise that results from the random motion of charge carriers in any resistive material. The mean-square thermal noise voltage, sometimes referred to as Johnson-Nyquist noise after those who first understood its origin,[10] is given by

$$\overline{V_n^2} = 4kTR\,\Delta f \quad [\text{V}^2] \tag{7.21}$$

where k is Boltzmann's constant, T is the temperature in kelvins (K), R is the electrical resistance, and Δf is the electrical bandwidth and is independent of frequency, or "white." The rms noise voltage in a unit bandwidth can be expressed as

$$V_n = (4kTR)^{1/2} \quad [\text{V}/\text{Hz}^{1/2}] \tag{7.22}$$

The noise current I_n in a unit bandwidth is obtained by dividing the open-circuit noise voltage by R:

$$i_n = (4kTR)^{1/2}/R = (4kT/R)^{1/2} \quad [\text{A}/\text{Hz}^{1/2}] \tag{7.23}$$

For a network having more than one resistor, each at a different temperature, the noise voltage is computed as if arising from each independently. For two resistors, R_1 and R_2, connected in series, the mean-square noise voltage adds:

$$\overline{V_n^2} = 4k(R_1 T_1 + R_2 T_2)\,\Delta f \quad [\text{V}^2] \tag{7.24}$$

For equal temperatures, the noise voltage is given by

$$V_n = (4kTR_{\text{eq}})^{1/2} \quad [\text{V}/\text{Hz}^{1/2}] \tag{7.25}$$

where R_{eq} is the equivalent series resistance.

For two resistors in parallel at temperatures T_1 and T_2, the mean-square noise currents add as

$$\overline{i_n^2} = 4k\left(\frac{T_1}{R_1} + \frac{T_2}{R_2}\right)\Delta f \quad [\text{A}^2] \tag{7.26}$$

For equal temperatures, the noise current is given by

$$i_n = \left(4kT\frac{R_1 + R_2}{R_1 R_2}\right)^{1/2} = \left(\frac{4kT}{R_{eq}}\right)^{1/2} \quad [A/Hz^{1/2}] \qquad (7.27)$$

where R_{eq} is the equivalent parallel resistance.

Other noise forms, often termed "excess noise," occur in certain kinds of resistive materials. For example, granular carbon resistors exhibit a form of excess noise proportional to the current known as "current noise." Current noise may mask the thermal noise. Generally, wire-wound and metal-film resistors are free of current noise.

Example 4: Find the thermal noise voltage for a 1-Hz bandwidth lead sulfide (PbS) detector.

Given: The detector exhibits a dark resistance of 10^6 ohms at 300 K.

Basic equations:

$$V_n = (4kTR)^{1/2} \quad [V/Hz^{1/2}] \qquad (7.22)$$

Assumptions: Thermal noise dominates over other forms of excess noise.

Solution: The thermal noise is given by

$$V_n = (4kTR)^{1/2} = \{4 \times 1.38 \times 10^{-23}[J/K] \times 300[K]$$
$$\times 10^6[ohms]\}^{1/2}$$
$$= 1.29 \times 10^{-7} \text{ V/Hz}^{1/2}$$

where $k = 1.38 \times 10^{-23}$ J/K.   ∎

7.5.2 Noise in Vacuum Tubes

The current in a vacuum tube, such as a multiplier phototube detector, exhibits fluctuations that arise from the fact that the charge e is not infinitesimal. Furthermore, the moving charges do not flow in regularly spaced intervals but in a random fashion. Thus, the current, which is the number of charge carriers n_τ times the charge e divided by the sample time τ, varies from time to time. The steady current, i_0, is defined as the *average* number of charge carriers, $\overline{n_\tau}$, times e divided by the sampling time τ and is independent of time:

$$i_0 = \overline{n_\tau}e/\tau = n_0 e \quad [A] \qquad (7.28)$$

where $n_0 = \overline{n_\tau}/\tau$ is the average number of charge carriers per unit time.

The statistical distribution of independent events, such as the passing of an electron, is governed by the Poisson distribution function. In this case the mean-square deviation in the current for a sample time τ is

$$\overline{\Delta i_\tau^2} = ei_0/\tau \quad [A^2] \qquad (7.29)$$

Equation (7.29) shows that the mean-square deviation of the current is proportional to the charge e and inversely proportional to the sampling time. Thus, the mean-square deviation is large for short sampling times.

The mean-square fluctuations in the current depend upon bandwidth, as for thermal noise, up to the limiting frequency, which is determined by the electron transit time T_{tr}. This is the time required for the electron to move between electrodes. For frequencies lower than $1/T_{tr}$, the mean square current is given by the Schottky equation

$$\overline{i^2} = 2ei_0\,\Delta f \quad [A^2] \tag{7.30}$$

and the rms noise current in unit bandwidth, referred to as "shot noise," is given by

$$i(\text{rms}) = \sqrt{2ei_0} \quad [A\ \text{rms}/Hz^{1/2}] \tag{7.31}$$

The Schottky equation[11] is not valid for space charge conditions such as in a thermionic diode but is valid for a multiplier phototube in which the motion of each electron is an independent event.

Example 5:   Find the anode dark-current shot noise for a 1-Hz bandwidth multiplier phototube.

Given: The average anode dark current i_a is 1×10^{-9} A, and the multiplier gain is 1×10^6.

Basic equation:

$$i(\text{rms}) = \sqrt{2ei_0} \quad [A\ \text{rms}/Hz^{1/2}] \tag{7.31}$$

where i_0 is the average cathode current.

Assumptions: The dark current results from thermionic emission only, and the multiplier does not contribute to the noise. Average cathode current is

$$i_0 = 10^{-9}/10^6 = 10^{-15} \text{ A}$$

Solution: The cathode shot noise $i_c(\text{rms})$ is given by

$$i_c(\text{rms}) = \sqrt{2ei_0} = 1.79 \times 10^{-17} \text{ A rms}/Hz^{1/2}$$

where $e = 1.6 \times 10^{-19}$ C, and the anode noise current is

$$i_a = i_c G = 1.79 \times 10^{-11} \text{ A rms}/Hz^{1/2} \qquad \blacksquare$$

7.5.3   Noise in Semiconductors

Noise in semiconductors is empirically found to exhibit three distinct regions of the spectrum[12] as shown in Fig. 7.6. At high frequencies the dominant noise is thermal noise as given by Eq. (7.21). At intermediate frequencies the noise is described in terms of a characteristic frequency and

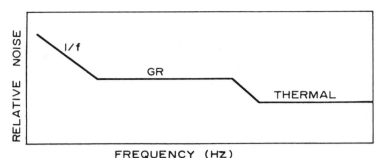

Figure 7.6 Noise in semiconductors.

is known as "generation-recombination" (gr) noise. At low frequencies the noise has a $1/f$ power dependence and is referred to as "current noise."

Generation-recombination noise is a major source of noise in semiconductors that is characterized by a power spectrum that is constant at intermediate frequencies but decreases rapidly beyond a characteristic frequency f. It has the form

$$\overline{i^2} = \frac{Ci^2}{1 + (f/f_1)^2} \qquad [\text{A}^2] \qquad (7.32)$$

where C is a constant. The characteristic frequency f_1 is related to the inverse of the carrier lifetime. The gr noise results from the statistical fluctuations in the concentration of carriers in a semiconductor and dominates over thermal noise at frequencies below f_1. In semiconductor photon detectors, gr noise is intimately associated with the random generation and recombination of charge carriers due to the random arrival of photons.

Current noise, which has a $1/f$ distribution, dominates over both gr and thermal noise at low frequencies. A number of sources have been suggested for this form of noise, including the electrical contacts of the wire leads and surface conditions on the semiconductor in which the conductivity of the material fluctuates and modulates the bias current. The latter results in a dependence of noise upon the level of bias current. Current noise, having a $1/f$ dependence upon frequency, cannot be reduced by reducing the bandwidth as can shot and thermal noise. However, the effects of $1/f$ noise can be minimized by limiting the spectrum to higher frequencies by chopping, not using a bias current, or other techniques.

7.5.4 Excess Noise

Electronic systems are subject to many forms of excess noise that can often be reduced by careful design. They include microphonics, pickup, and cross talk. Microphonic noise is caused by modulation of interelectrode capacitance as a result of mechanical vibrations. Current flows into and out of device terminals when the magnitude of the interelectrode capacitance is modulated. This occurs, for example, when a moving light-chopper blade passes over a detector without proper shielding. The resultant currents induced in the detector cannot be distinguished from chopped radiant flux.

7.6   COHERENT RECTIFICATION

The coherent rectifier was first utilized for microwave radiometry by Dicky.[13] The basic idea has been applied to many other fields and applications; for example, the "lock-in" and phase-lock-loop amplifiers. The technique wherein the incoming radiation is chopped, ac-amplified, and subsequently coherently detected to produce a dc output proportional to the incoming signal has also been referred to as "phase-sensitive detection."[14]

As previously indicated, most electro-optical systems employ light choppers to transform the information-frequency spectrum of the radiant signal to the chopping frequency in order to provide the most favorable detector-noise characteristics. This permits the use of ac-coupled amplifiers that block the dc and low-frequency drift components of the detector output. This is illustrated in Fig. 7.5, where f_m is the maximum modulation frequency.

The ac signal consists of a carrier at the frequency f_c that is proportional to the dc component of the radiant signal and sidebands $f_c \pm f_m$ that correspond to the radiant signal modulation f_m. Negative modulation frequencies do not exist, but it is mathematically convenient to include $-f_m$ in the model. In this context the dc component of the radiant signal can be viewed as the information carrier in electro-optical systems. The average value of the carrier is never zero as in more conventional communication systems.

One consequence of the double sidebands in the transformed information spectrum is that the electrical ac bandwidth must be twice that which is necessary for a dc radiometer. Thus, choppers are not employed unless significant improvement in the noise spectrum is realized at f_c compared to dc.

"Dc restoration" is a term used to describe a process in which the sensor output is made to reproduce the waveform of the flux beam. The process utilizes "coherent rectification" and a low-pass filter that establishes the information/noise bandwidth. A functional flow diagram of a radiometer that utilizes dc restoration is given in Fig. 2.2.

Incoherent rectifiers introduce nonlinearity so the output signal is not proportional to the input flux. Such a rectifier is called "incoherent" because the output does not depend upon the phase of the input signal. In addition, the noise spectrum is modified by the nonlinear rectifier as a result of the production of cross products in the output.

A *coherent rectifier* is one whose output is proportional to the amplitude of the ac input regardless of its magnitude. The simplest form of this type of rectifier is a switch that causes the output to be reversed every half-cycle of the signal frequency, producing full-wave rectification.

A very important aspect of coherent rectification results from the fact that periodic reversal of the phase of random noise has no effect upon its statistics and thus the noise mean and variance remain unchanged. The mean value of the noise is zero, and the standard deviation can be reduced to arbitrarily low values by filtering.

The action of the coherent rectifier can be represented as the *product* of the ac signal and a square wave; the resultant cross products are given by Eq. (7.7). In the case of coherent rectification, the switching frequency and phase of the rectifier are the same as that of the ac signal; this arrangement has also been referred to as a "homodyne" amplifier.[15] Mixing the sampled frequency f_c with a locally generated switching function of the same frequency and phase yields the sum and difference frequencies—the difference is dc—proportional to the sampled mean flux. The higher-order components are removed by the low-pass filter.

Coherent rectification restores the original sampled signal. The lower sideband is folded over onto the upper sideband (since negative frequencies do not exist), resulting in a reduction in the effective bandwidth by a factor of 2. This results in two important considerations: First, the ac bandwidth must be at least a factor of 2 greater than the final low-pass filter so that neither sideband is lost; second, the foldover effect of the rectifier increases the noise.[16] Provided that the noise in the sidebands is statistically independent, the resultant rms noise in the output increases by the factor $\sqrt{2}$.

The fundamental bandwidth requirements for a system design arise from considerations of the information bandwidth and the sampling theorem. The ac circuit bandwidth can be appropriately limited, in principle, using a narrow-band filter. However, it is difficult to design bandpass filters that are relatively narrow and stable. Consequently, the ac bandwidth is generally set somewhat greater than $2f_m$. Then the fundamental limiting information-noise bandwidth is determined by the low-pass filter.

For the special case where the radiant signal is stationary in time (no modulation), the break frequency of the low-pass filter can be arbitrarily reduced, and the noise tends to its mean value, zero. Thus it is possible, using coherent rectification techniques, to extract useful signals from the ac output for which the signal-to-noise ratio is much less than unity. The use of the coherent detector permits the selection of a narrow band of noise about the sample frequency *after detection*, ignoring all noise outside this band.

A narrow-band filter can be used with incoherent rectification also, but noise from all parts of the spectrum contribute to the noise in the filter band because of the product mixing of these components in the nonlinear diode used in such rectifiers.

Bandpass filters are generally used in the ac amplifiers of coherent systems to prevent overdriving of the output stages with wide-band noise and to help reduce the effect of power supply ripple and other environmental forms of noise pickup.

Reversal of the phase of the switching frequency by 180° results in negative output of the coherent detector. In some special cases the phase is adjusted to $\pm 90°$, which produces a null output.

EXERCISES

1. A detector noise voltage is measured at 1.5×10^{-6} V rms with a 1000-Hz noise bandwidth at a temperature of 300 K. Find the thermal noise equivalent resistance.

2. For detector R_d and bias resistor R_b in Fig. E7.2, where $R_b = 10^7$ ohms, $R_d = 10^8$ ohms, and $T = 300$ K, find the limiting thermal noise voltage in units of $V/Hz^{1/2}$.

3. The rise time T_r is defined as the time required for a system to respond from 10 to 90% to an input step function. It is given by $T_r = 0.35/f_2$, where f_2 is the upper-frequency break point. Find the percent output [Eq. (7.1)] obtained for an ideal square-wave input pulse width equal to the rise time, i.e., $T_d = T_r$.

4. Derive the output of a mixer for which the sample function is a sine wave $(\sin \omega_s t)$ and the input (sampled function) is a sine-wave-modulated dc component $(a + b \sin \omega_m t)$. Given $f_m = 5$ Hz and $f_s = 100$ Hz. Plot the results as in Fig. 7.5.

5. A continuous signal contains the sum of two sine-wave frequencies, $f_1 = 50$ Hz and $f_2 = 75$ Hz. Suppose it is sampled with a 100-Hz sine-wave sampling function. Find all the frequencies in the mixer output, and plot each component as in Fig. 7.5.

6. A continuous signal contains a modulation component (sine wave) at 100 Hz. It is sampled with a square-wave $\Sigma b_n \sin(n\omega_s t)$ sampling function at 300 Hz. The system specifications require that the higher-order products of the mixer be filtered 60 dB below the 100-Hz signal with a bandpass filter. Find the minimum filter attenuation in dB/octave or dB/decade to achieve the specified filtering.

7. A sequential spectrometer has a scan rate $S = 0.5$ Hz and contains $N = 50$ resolution elements. Find the dwell time T_d. The dwell time is equal to the integration time in this case. Find the frequency response if the system is to achieve the magnitude of the resolution element within 1%.

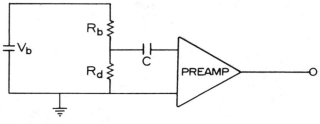

Figure E7.2

8. A sequential spectrometer has a scan rate of 0.5 Hz and $N = 50$ resolution elements designed for 99% magnitude accuracy.

  (*a*) Find the maximum frequency in the spectrometer output (assuming the source is not modulated, i.e., $f_m = 0$).

  (*b*) Find the minimum chopping frequency to satisfy the sampling theorem.

9. A 741 op-amp is used with ± 15-V supplies. The output noise voltage is 12 μV rms. Find the approximate dynamic range of the output.

10. A multiplier phototube has an anode dark current i_0 (average current) of 10^{-9} A and a multiplier gain of 1×10^6. Suppose the ac electrical noise bandwidth is 10 Hz; what is the anode rms noise current?

11. A sequential spectrometer is to be designed to measure $N = 100$ resolution elements. Determine the scan rate and chopper frequency if the highest radiant source modulation frequency is 5 Hz. *Hint:* Use the sampling theorem, and assume ideal filters.

12. Find the limiting thermal noise rms voltage, in $V/Hz^{1/2}$, for two resistors connected in parallel. $R_1 = 1 \times 10^{10}$ ohms, $R_2 = 5 \times 10^9$ ohms, and the temperature is 300 K.

13. For a pulse code modulation (PCM) fiber optics link with the following specifications, find the link length possible.

15 channels multiplexed (15 words/frame)
Information bandwidth = 3 kHz
Accuracy: > 99% (12 bits)
Word synchronization: 2 bits
Frame synchronization: 1 word
Source: LED
    Output flux: 150 mW at 6300 A
    Rise time: 100 ns
Link loss: 15 dB/km
Coupler losses: 3 dB each (2 minimum)
SNR: 20 dB
Detector: Multiplier phototube
    Rise time 8 ns
    NEP: 1×10^{-14} $W/Hz^{1/2}$
    Low-pass filter cutoff: $\frac{1}{2}$ sample rate

Hint: It is not necessary to have amplitude accuracy in a digital system; hence, $f_2 = \frac{1}{2}$ bit rate is an empirical rule.

REFERENCES

1. E. J. Kelley, D. H. Lyons, and W. L. Root, "The Sensitivity of Radiometric Measurements," *J. Soc. Indust. Appl. Matt.*, **11**, 235 (1963).
2. W. W. Harman, *Principles of the Statistical Theory of Communication*, McGraw-Hill, N.Y., 1963, p. 217.
3. L. A. Wainstein and V. D. Zubakov, *Extraction of Signals from Noise*, Prentice-Hall, Englewood Cliffs, NJ, 1962, p. 39.
4. W. W. Harman, *Principles of the Statistical Theory of Communication*, McGraw-Hill, N.Y., 1963, p. 28.

5. E. V. Loewenstein, "Fourier Spectroscopy: An Introduction," *Aspen Int. Conf. Fourier Spectry.*, Spec. Rep. N. 114, AFCRL-71-0019, Air Force Cambridge Res. Lab., L. G. Hanscom Field, Bedford, MA, 1970, p. 4.

6. H. L. Stiltz, Ed., *Aerospace Telemetry*, Prentice-Hall, Englewood Cliffs, NJ, 1961, p. 84.

7. *Ibid.*, p. 188.

8. *Ibid.*, p. 190.

9. *Ibid.*, p. 85.

10. P. W. Kruse et al., *Elements of Infrared Technology: Generation, Transmission, and Detection*, Wiley, New York, 1962, p. 236.

11. R. A. Smith et al., *The Detection and Measurement of Infra-red Radiation*, Oxford University Press, London, 1957, p. 191.

12. P. W. Kruse et al., *Elements of Infrared Technology: Generation, Transmission, and Detection*, Wiley, New York, 1962, p. 251.

13. R. H. Dicke, "The Measurement of Thermal Radiation at Microwave Frequencies," *Rev. Sci. Instrum.*, **17**, 260–275 (1946).

14. R. A. Smith et al., *The Detection and Measurement of Infra-red Radiation*, Oxford University Press, London, 1957, pp. 236–242.

15. *Ibid.*, p. 431.

16. *Ibid.*, p. 236.

chapter 8

The Radiometric Performance Equation

8.1 INTRODUCTION

The radiometric performance equation is central to the approach to system design given in this text. Its development in this chapter follows the development of the "measurement equation" given in the National Bureau of Standards "Self-Study Manual."[1,2] A general expression of the equation is written first; and then by appropriate assumptions and simplifications the general equation is rewritten for two illustrative systems. The examples given here deal only with the spatial, spectral, and temporal domains. Polarization is neglected. The two measurement configurations are (1) the extended-area source and (2) the distant small-area source. In each case, the detector is characterized in terms of the detectivity.

This chapter also includes (1) a general discussion that summarizes the important features of the radiometric performance equation; (2) a set of orderly steps for writing the equation; and finally, (3) a summary of the limitations of the radiometric performance equation.

8.2 THE RADIOMETRIC PERFORMANCE EQUATION

The radiometric performance equation provides a criterion for the optimum design of electro-optical systems that is based on the premise that an overriding requirement for all systems designed for information processing is adequate signal-to-noise ratio (SNR). In the radiometric analysis of the feasibility study (Sec. 2.5.1), top priority is given to the signal-to-noise ratio.

Then, depending upon the specific application, spatial or spectral resolution, speed of response, etc., might be important in trade-off studies.

A general form of the equation was given in Sec. 2.6 as the ratio of the effective flux Φ_{eff} to the noise equivalent flux (NEF).

$$\text{SNR} = \frac{\Phi_{\text{eff}}}{\text{NEF}} \tag{8.1}$$

where Φ_{eff} is the effective flux, defined as the magnitude of flux incident upon the sensor detector that is effective in evoking a response in the sensor output as given in Sec. 5.10. The NEF is defined as the change in flux tht produces an average change in the output signal equal to the root-mean-square (rms) noise in the output signal.[3]

Table 8.1 lists the system and subsystem figures of merit with the corresponding symbols and units. From the list of alternative subsystem figures of merit, those must be selected for inclusion in the equation that provide for optimization of system performance goals while simultaneously maximizing the signal-to-noise ratio.

It is important to remember that polarization is being ignored in the development given here. In addition, the effects of nonlinearity and hysteresis are being set aside although they are often a consideration in the design of real systems.

Other assumptions about the system must be made before the performance equation can be written. For example, the appropriate figure of merit for the detector depends upon the limiting noise mechanism and the detector responsivity. For purposes of illustration, it is assumed that the detector can be characterized in terms of the detectivity.

Equation (8.1) can now be expanded in terms of subsystem figures of merit for a measurement configuration as illustrated in Fig. 2.3, using Eqs. (5.28) for the effective flux upon a detector and (6.6) for the detector NEP in terms of D^* and correcting for dc restoration as follows:

$$\text{SNR} = \frac{D^*(\lambda)\tau_p\tau_e}{\beta(2A_d f_2)^{1/2}} \int_\theta\int_\Phi\int_\lambda \Phi_s(\lambda, \theta, \Phi)\mathscr{R}(\lambda)\, d\theta\, d\phi\, d\lambda \tag{8.2}$$

where $D^*(\lambda)$ is the detectivity, in cm $\text{Hz}^{1/2}/\text{W}$

       τ_e is the optical efficiency (unitless)

       β is the chopping factor (unitless)

      $\sqrt{2}$ is the phase detection noise factor (unitless)

      A_d is the detector area, in cm^2

      f_2 is the electrical (low-pass) noise filter cutoff frequency, in $s - 1$

   $\Phi_s(\lambda)$ is the source spectral flux, in $\phi/\mu\text{m}$

   $\mathscr{R}(\lambda)$ is the sensor relative spectral response function

     Φ is the polar angle

     ϕ is the azimuthal angle

     λ is the wavelength, in μm

Table 8.1 SYSTEM AND SUBSYSTEM FIGURES OF MERIT

Term	Symbol	Unit
Source		
Sterance [radiance]	L	W cm^{-2} sr^{-1}
Areance [exitance]	M	W/cm^2
Pointance [intensity]	I	W/sr
Path		
Path transmittance	τ_p	—
Receiver		
Areance [irradiance]	E	ϕ/cm^2
Aperture area	A_c	cm^2
Spatial		
Throughput (general)	$A\Omega$	cm^2 sr
Relative aperture	F	—
Field of view (linear)	θ	deg (°), rad
Field of view (solid)	Ω	sr
Spectral		
Free spectral range	$\lambda_1 - \lambda_2$	μm
Bandwidth	$\Delta\lambda$	μm
Resolving power	$\delta\lambda$	μm
Polarization		
Stokes parameters		
Optical		
Optical efficiency	τ_e	—
Chopping factor	β	—
Detectors		
Noise equivalent power	NEP	W
Responsivity	$\mathscr{R}$	V/W
Detectivity	D^*	cm Hz$^{1/2}$/W
Electronics		
Noise bandwidth	Δf	Hz (s^{-1})
Time constant	T_c	s
Rise time	T_r	s
Noise voltage	V_n	V rms
System		
Noise equivalent power	NEP	W
Noise equivalent sterance	NES	ϕ cm^{-2} sr^{-1}
[Noise equivalent radiance]	[NER]	W cm^{-2} sr^{-1}
Noise equivalent spectral sterance	NESS	ϕ cm^{-2} sr^{-1} μm^{-1}
[Noise equivalent spectral radiance]	[NESR]	W cm^{-2} sr^{-1}
Noise equivalent flux density	NEFD	W/cm^2
Noise equivalent spectral flux density	NESFD	W cm^{-2} μm^{-1}

8.2.1   The Extended-Area Source

The radiometric performance equation must be written in a form containing the desired radiometric quantity to characterize the source. It is written most appropriately for the extended-area source in terms of the sterance [radiance] as follows:

$$\text{SNR} = \frac{D^*(\lambda)\tau_p\tau_e}{\beta(2A_df_2)^{1/2}} \int_{A_s}\int_{\omega_s}\int_{\lambda} L_s(\lambda)\mathscr{R}(\lambda)\cos\theta\, dA_s\, d\lambda\, d\omega_s \quad (8.3)$$

where $L_s(\lambda)$ is the source sterance [radiance] in W cm^{-2} sr^{-1} μm^{-1} and A_s is the source area in cm^2.

Equation (8.3) can be integrated based upon a number of simplifying assumptions:[4]

1. The source sterance [radiance] is uniform over the source area A_s.
2. The source sterance [radiance] is uniform over the system spectral response function.
3. The solid angle is independent of the source area.
4. The path is contained in a homogeneous medium where the index of refraction exhibits unity value.

Then Eq. (8.3) can be written as

$$\text{SNR} = \frac{L_s(\lambda)\, \Delta\lambda\, D^*(\lambda)\tau_e\tau_p A_{\text{fs}}\pi}{\beta(2A_df_2)^{1/2}4F^2} \quad (8.4)$$

where, by the invariance theorem,

$$\Omega_s A_s = \Omega_c A_c = A_{\text{fs}}\pi/4F^2 \quad (8.5)$$

and where $\int dA_s = A_s$, the source area, in cm^2

$\int \cos\theta\, d\omega_s = \Omega_s$, source solid angle, in sr

$\int L_s(\lambda)\mathscr{R}(\lambda)\, d\lambda = L_s(\lambda)\, \Delta\lambda$, the effective flux, in W cm^{-2} sr^{-1}

For the case where the detector serves as the field stop, $A_{\text{fs}} = A_d$, as illustrated in Fig 5.3, Eq. (8.4) can be written as

$$\text{SNR} = \frac{L_s(\lambda)\, \Delta\lambda D^*(\lambda)\tau_e\tau_p\pi(A_d/2f_2)^{1/2}}{4F^2\beta} \quad (8.6)$$

Equation (8.6) provides for system optimization and sensitivity analysis:[5] In consideration of the most obvious parameters, the SNR is maximized by maximizing the source spectral sterance [radiance] $L_s(\lambda)$, the detectivity D^*, the optical efficiency τ_e, and the path transmittance τ_p. The effect of the magnitude of the bandwidth $\Delta\lambda$, the detector area A_d, the relative aperture F, and the electrical noise frequency f_2 warrants more careful consideration.

Equation (8.6) illustrates that the SNR is more sensitive to the relative aperture, or f-number, than any other parameter. Lower f-number corresponds to "faster optics" and to increased throughput and SNR. However, Eq. (8.6) does not contain an explicit term for the sensor field of view. It is true that the SNR is independent of field of view in the case of a uniform extended-area source; this follows from the fact that field of view can be traded off for aperture area.

The SNR is not very sensitive to increases in the throughput that are accomplished by increasing the detector field-stop area. This results from the fact that for detectors most appropriately characterized with D^*, the noise increases with the square root of detector area; hence, NEP is proportional to the square root of detector area.

The conclusions drawn from Eq. (8.6) may not be valid for different noise-limiting cases. For example, extrinsic detectors operated at cryogenic temperatures and shielded from any background may not produce measurable noise.[6] In this case the limiting noise originates in the preamplifier, and Eq. (6.5) is a more appropriate expression of the detector NEP. The preamplifier noise is often directly proportional to the detector capacitance and to detector area. In this case the SNR is more or less independent of detector area.

The electrical noise bandwidth also determines the system information bandwidth. Equation (8.6) shows that SNR is inversely proportional to the square root of bandwidth. This means that NEP is not very sensitive to noise bandwidth and that information bandwidth can often take precedent over SNR.

The SNR is directly proportional to the bandwidth $\Delta\lambda$, and a trade-off is necessary between SNR and spectral resolution. This becomes a serious limiting factor for high-resolution spectrometer sensors. High spectral resolution has implications for noise bandwidth also. In most applications, the time available for a complete scan is fixed, and increasing the resolution increases the bandwidth requirement further, compromising the SNR.

8.2.2 Distant Small-Area Source

In this case the radiometric performance equation is to be written in terms of a distant small area or "point source." It is most appropriate to characterize a nonresolvable point source in terms of the pointance [intensity] I in W/sr, using Eqs. (5.32) and (6.6) and correcting for dc restoration as follows:

$$\text{SNR} = \frac{I_s(\lambda)\,\Delta\lambda\,D^*(\lambda)\tau_e\tau_p\pi\left(A_d/2f_2\right)^{1/2}}{4F^2\beta\Omega_c s^2} \tag{8.7}$$

where $I_s(\lambda)$ is the source spectral pointance [intensity] in units of

W sr^{-2} μm^{-1} and the factor $\sqrt{2}$ is included for coherent rectification (dc restoration). Equation (8.7) is useful for optimization and sensitivity analysis for point sources where there exists a particular field-of-view requirement. The conclusions are similar to those given in Sec. 8.2.1 for the extended-area source in that the SNR is proportional to the square root of detector area and inversely proportional to the f-number squared.

8.3   RADIOMETRIC PERFORMANCE EQUATION AND ITS USE

The SNR equation has been introduced by writing it for specific cases: the extended-area source and the point source. In each case the effective flux was described in terms of integrals of the source and the sensor elements. The detector NEP was described in terms of the detectivity D^*. However, D^* is appropriate for only one class of detectors. Actually, what is important is the limiting noise mechanism and its relationship to the detector area or throughput.

The integrals in the radiometric performance equation can be solved by employing certain simplifying assumptions as indicated above.[4] In principle, it is always possible to separate the variables when the spectral flux is uniform with wavelength and area, and when the solid angle and area are independent. Configuration factors can be used when area and solid angle are not independent.

The following three steps will assist the designer in making use of the radiometric performance equation to accomplish a feasibility study.

Step 1   The choice of a detector and its operating mode are always at the heart of the design of an electro-optical sensor. This choice must be based upon the wavelength, flux level, speed of response, and signal-to-noise ratio required. A first cut on this selection process is made by solving Eq. (5.26), written in a form appropriate for the source, to obtain the effective flux on the detector. Then, using Eq. (6.3) and the required SNR, find the detector D^*. Figure 8.1 gives D^* as a function of wavelength for a set of different classes of detectors. Using Fig. 8.1 and similar information, it is possible to determine the operating conditions for the appropriate detector.

The easiest way to find the effective flux is to prepare a diagram showing the source A_s, the separating distance s, and the sensor collecting aperture A_c. Use the form of Eq. (5.28) for an extended-area source. The throughput can be expressed in terms of the source ($A_s\Omega_s$), the entrance aperture ($A_c\Omega_c$), or the focal plane ($A_{fs}\pi/4F^2$). Use the form of Eq. (5.32) for the distant small-area (point) source. Generally, the f-number, F, is the most useful parameter in determining the sensor throughput.

Example 1:   Select a detector from Fig. 8.1 to satisfy the requirements listed below. The system is a narrow-band radiometer that utilizes a light chopper and coherent rectification (dc restoration).

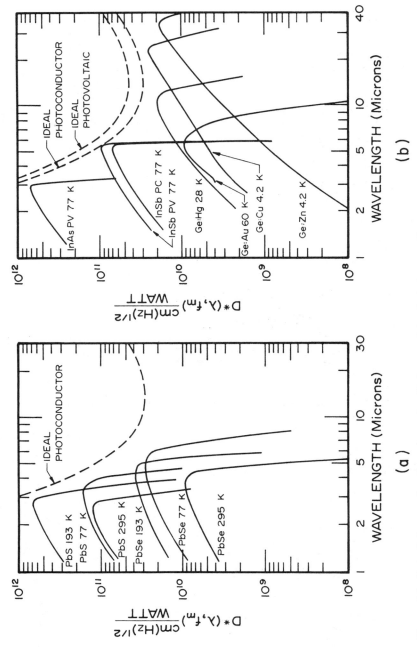

Figure 8.1 Detectivity as a function of wavelength for a variety of photon-counting detectors. Note: Values given are for a 295-K background and π sr field of view. (*Courtesy of Santa Barbara Research Center.*)

Given:

Point-source target pointance [intensity] $I_s(\lambda) = 48$ W sr^{-1} μm^{-1} at 3.5 μm

Optical bandwidth $\Delta\lambda = 0.27$ μm

System optical efficiency $\tau_e = 0.3$

Path transmittance $\tau_p = 0.5$

f-number $F = 3.0$

Detector diameter $= 1.0$ mm

SNR $= 100$ (minimum)

Chopping factor $\beta = 3$

Information-noise bandwidth $\Delta f = 0.1$ Hz

Field-of-view solid angle $\Omega_c = 10^{-4}$ sr

Target distance $s = 1$ km

Basic equations:

$$\Phi_{eff} = \frac{I_s(\lambda)\,\Delta\lambda\,\tau_p\tau_e A_d \pi}{4F^2\Omega_c\beta s^2} \qquad (5.32)$$

$$D^*(\lambda) = (A_d\,\Delta f)^{1/2}/\text{NEP} \qquad [\text{cm Hz}^{1/2}/\text{W}] \qquad (6.3)$$

$$\text{SNR} = \frac{I_s(\lambda)\,\Delta\lambda\,D^*(\lambda)\tau_e\tau_p\pi(A_d/2f_2)^{1/2}}{4F^2\beta\Omega_c s^2} \qquad (8.7)$$

Assumptions: The bandwidth and field of view are small enough to permit the integration of Eq. (5.26) into the form given here, namely, Eq. (5.32).

Solution: Examination of Fig. 8.1 shows that a large number of detectors respond at 3.5 μm. The required D^* can be calculated from the above-referenced equations and then a selection made based upon the simplest possible detector implementation.

The effective flux is given by Eq. (5.32):

$$\Phi_{eff} = \frac{48 \times 0.27 \times 0.5 \times 0.3 \times 7.85 \times 10^{-3} \times \pi}{4 \times 3^2 \times 3 \times 1 \times 10^{-4} \times 1 \times 10^{10}} = 4.44 \times 10^{-10} \text{ W}$$

For SNR $= 100$, the NEP must be

$$\text{NEP} = 4.4 \times 10^{-10}/100 = 4.4 \times 10^{-12} \text{ W}$$

The required detectivity is given by Eq. (6.3), where the factor $\sqrt{2}$ is added for coherent rectification:

$$D^*(\lambda = 3.5 \ \mu\text{m}) = \frac{(2 \times 7.85 \times 10^{-3} \times 0.1)^{1/2}}{4.4 \times 10^{-12}}$$

$$= 9.0 \times 10^9 \text{ cm Hz}^{1/2}/\text{W}$$

Figure 8.1 shows that InSb has a D^* of 5 to 7×10^{10} cm $Hz^{1/2}/W$ at 3.5 μm; moreover, this detector must be operated at cryogenic temperatures, which is inconvenient and expensive. However, PbSe operated at ambient temperatures (295 K) has the required D^* of about 1×10^{10} cm $Hz^{1/2}/W$ and will satisfy the requirements given above; it represents the most economical approach to the design. The solution can be verified by solving Eq. (8.7) for the SNR. ∎

Step 2 Based upon the choice of a detector and with the knowledge of the dominant noise mechanism, it is now possible to write an expression for the detector NEP. Its form depends upon whether or not the system is dc coupled or chopped and the dominant noise mechanism. Detectivity D^* is appropriate for one class of detectors. Detectors operated at reduced backgrounds require that the NEP be expressed in terms of the responsivity and the limiting noise. Whenever the noise is a function of detector area, the effective flux and NEP are interdependent.

Step 3 It should now be possible to write the SNR radiometric performance equation by combining the expressions for effective flux and NEP in accordance with Eq. (8.1). Detector area factors common to the effective flux and the NEP can be combined. In this form, the radiometric performance equation provides for optimization, sensitivity analysis, and trade-offs.

Example 2: Select a detector from Fig. 8.1 to satisfy the requirements listed below. The system is a narrow-band radiometer that utilizes a light chopper and coherent rectification (dc restoration).

Given:

Extended-area source sterance [radiance] $L(\lambda) = 1 \times 10^{-6}$ W cm^{-2} sr^{-1} μm^{-1} at 9.6 μm

Optical bandwidth $\Delta\lambda = 0.8$ μm

System optical efficiency $\tau_e = 0.2$

Path transmittance $\tau_p = 1.0$

f-number $F = 2.0$

Detector diameter = 1.0 mm

SNR = 10 (minimum)

Chopping factor $\beta = 3$

Information-noise bandwidth $f_2 = 1$ Hz

Basic equation:

$$\text{SNR} = \frac{L_s(\lambda) \, \Delta\lambda \, D^*(\lambda) \, \tau_e \tau_p \pi \left(A_d / 2 f_s \right)^{1/2}}{4 F^2 \beta} \qquad (8.6)$$

Assumptions: The bandwidth and field of view are small enough to permit the integration of Eq. (5.27) into the form given here.

Solution: Examination of Fig. 8.1 leads to the conclusion that the best detectivity at 10 μm is obtained using GeHg at 28 K, where the detectivity is given as 2×10^{10} cm $Hz^{1/2}/W$.

The signal-to-noise ratio is

$$\mathrm{SNR} = \frac{1 \times 10^{-6} \times 0.8 \times 2 \times 10^{10} \times 0.2 \times \pi \times (7.85 \times 10^{-3})^{1/2}}{4 \times 2^2 \times 3 \times \sqrt{2}}$$

$$= 13.1$$

Note: Although the GeHg detector satisfies the SNR requirement, it requires operation at 28 K. The above operating conditions—π sr field of view to a 295 K background—are referred to as BLIP (background limited infrared photodetector), and the limiting noise is photon noise originating in the incident background.   ■

8.4   LIMITATIONS TO THE RADIOMETRIC PERFORMANCE EQUATION

Most of the limitations to the radiometric performance equation have already been discussed. They are summarized here. The equation does not include the effects of:

1. Polarization
2. Environmental effects, other than detector operating temperture, such as vibration, and electric and magnetic fields
3. Diffraction and scattering effects (The discussion has been primarily based upon radiation geometry.)
4. Nonlinear response

In addition, the very important considerations of out-of-band rejection and off-axis rejection have not been considered. However, system designs for high out-of-band or off-axis rejection usually have only a second-order effect upon the system SNR. Thus, the solution of the radiometric performance equation, as outlined above, is generally adequate for a feasibility study.

The ability to correctly discern the dominant noise mechanism in a given radiometric system is the most serious limitation of the textual development to this point. Noise forms that must be considered include thermal, or Johnson, noise in resistive elements, which is a function of operating temperture; shot noise in photoemissive devices, which is a function of temperature and average current; preamplifier noise, which is a function of input capacitance and frequency; and photon noise, which is a function of the uncoded background incident upon the detector. The dominant noise can be found using the methods given in Chap. 18.

A method to solve the radiometric performance equation with respect to the noise problem is to solve for a limiting case. The most fundamental noise limitation is photon noise: In this case the detectivity D^* is the appropriate parameter to use. In the event that the detector is shielded from any significant background radiation, D^* cannot be used.[6] Then the next most fundamental limit is thermal noise generated in the detector. The detector bias resistor thermal noise must also be considered; in general, the thermal noise is given by

$$V_n(\text{rms}) = (4kTR\,\Delta f)^{1/2} \tag{8.8}$$

where R is either the bias resistance or the detector resistance. Generally, the preamplifier noise dominates for wide-band systems.

EXERCISES

1. Verify the solution of Example 1 by solving the appropriate radiometric performance equation for the SNR.

2. Find the SNR for a narrow-band radiometer that employs a light chopper and coherent rectification (dc restoration) for the following specifications:

   Source sterance [radiance] $L_s(\lambda) = 1 \times 10^{-6}$ W cm^{-2} sr^{-1} μm^{-1}
   Optical bandwidth $\Delta\lambda = 1.0$ μm
   Detector $D^*(\lambda_p) = 1 \times 10^{10}$ (PbSe at 295 K)
   Optical efficiency $\tau_e = 0.7$
   Path transmittance $\tau_p = 1.0$
   Detector diameter $= 1.0$ mm
   f-number $F = 3$
   Chopping factor $\beta = 3$
   Information-noise bandwidth $= 1.0$ Hz

3. Write a general equation similar to that of Eq. (8.6) for the SNR of a narrow-band radiometer system for which the InSb detector is operated under low-background conditions at 77 K. Assume an extended-area source and coherent rectification. *Hint:* Assume the system is detector resistance termal noise limited.

4. Find the SNR for a narrow-band radiometer that utilizes a light chopper and coherent rectification (dc restoration) for the following:

   Detector InSb at 77 K, low-background
   Detector resistance $R_d = 2 \times 10^8$ ohms
   Detector diameter $= 1$ mm
   Detector-preamplifier responsivity $\mathscr{R}(\lambda) = 1 \times 10^9$ V/W at 5.0 μm
   Optical bandwidth $\Delta\lambda = 0.8$ μm
   Source sterance [radiance] $L_s(\lambda) = 1 \times 10^{-11}$ W cm^{-2} sr^{-1} μm^{-1}
   Optical efficiency $\tau_e = 0.5$
   Path transmittance $\tau_p = 1.0$
   Chopping factor $\beta = 3$
   f-number $F = 2$
   Information-noise bandwidth $= 1.0$ Hz

   Hint: Assume the limiting noise is detector resistance thermal noise (see Exercise 3).

5. Write a general equation similar to that of Eq. (8.6) for the SNR of a narrow-band radiometer system given that the source is characterized in terms of the incident areance [irradiance] E in W/cm². Express the system throughput in terms of relative aperture F and detector area A_d. Include terms for chopping factor, optical efficiency, and coherent rectification (dc restoration).

6. Write a general equation similar to that of Eq. (8.6) for the SNR of a narrow-band radiometer system given that the source is a distant small area (does not fill the field of view) of area A_s, is at a distance s, and is radiating L_s W cm⁻² sr⁻¹. Express the SNR in terms of sensor field of view Ω_c, detector area A_d, and relative aperture F. Include terms for path losses, optical efficiency, chopping factor, and coherent rectification (dc restoration).

7. Find the detector D^* required to achieve the following specifications:

> Radiometer system using chopper and coherent rectification
> Source areance [radiance] $L_s = 1 \times 10^{-8}$ W cm⁻² sr⁻¹
> Optical efficiency $\tau_e = 1$
> Path transmittance $\tau_p = 1$
> Detector diameter = 1 mm
> Information/noise bandwidth $f_2 = 12$ Hz
> Signal-to-noise ratio = 100

REFERENCES

1. F. E. Nicodemus, Ed., "Self-Study Manual on Optical Radiation Measurements. Part 1—Concepts," *Natl. Bur. Stand. (U.S.)*, *Tech. Note* No. 910-1, 1976, pp. 4–7.

2. F. E. Nicodemus, Ed., "Self-Study Manual on Optical Radiation Measurements. Part 1—Concepts," *Natl. Bur. Stand. (U.S)*, *Tech. Note* No. 910-2, 1978, pp. 58–92.

3. E. J. Kelley, D. H. Lyons, and W. L. Root, "The Sensitivity of Radiometric Measurements," *J. Soc. Indust. Appl. Math.*, **11**, 235–257 (1963).

4. F. E. Nicodemus, Ed., "Self-Study Manual on Optical Radiation Measurements. Part 1—Concepts," *Natl. Bur. Stand. (U.S.)*, *Tech. Note* No. 910-2, 1978, p. 86.

5. W. L. Wolfe and George J. Zissis, Eds., *The Infrared Handbook*, Office of Naval Research, Dept. of the Navy, Washington, DC, 1978, p. 19-5.

6. W. L. Wolfe, *Handbook of Military Infrared Technology*, Office of Naval Research, Dept. of the Navy, Washington, DC, 1965, p. 516.

chapter 9

Feasibility Study — An Example

9.1 INTRODUCTION

The purpose of this chapter is to present a practical example of a feasibility study. Such a presentation might be based upon any one of a great number of applications. The example given here represents an instrumentation application in which a rocketborne sensor is designed to obtain airglow data measured on the earth limb (outer edge of earth) from a rocket platform above the atmosphere. Unfortunately, any single example fails to illustrate all the important aspects of a feasibility study. However, the basic principles used in this example are common to most systems.

9.2 PROGRAM OBJECTIVES

This feasibility study is for a multispectral scanning radiometer to investigate several aurorally enhanced emitters including hydroxyl (OH) and carbon dioxide (CO_2) emissions in the upper atmosphere. The measurement program is referred to as ELIAS, for Earth Limb Infrared Atmospheric Structure.[1] The instrument, equipped with a well-baffled telescope, is directed at the earth limb as illustrated in Fig. 9.1. Spatial variations in the airglow are to be measured through the use of a focal plane array and an internal scanning mirror.

The OH fundamental ($\Delta\nu = 1$) emission occurs at 2.7 μm and is observed as a 10-km thick layer that exhibits a peak at 87 km in the lower

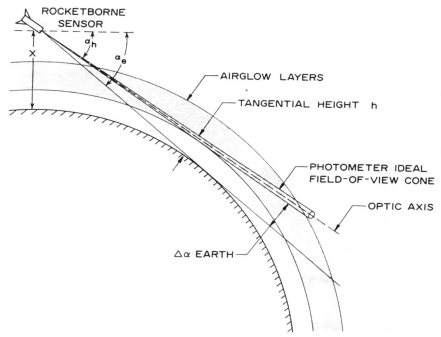

Figure 9.1 Illustration of the earth-limb mode to obtain the vertical distribution of an atmospheric emission species.

thermosphere.[2] The CO_2 $\nu(3)$ emission occurs at 4.3 μm with a peak below 45 km and decreases logarithmically to above 100 km.[3] The N_2^+ emission at 3914 Å provides a monitor of the auroral activity resulting from electron deposition in the upper atmosphere from solar events.[4] Such energy deposition is known to be a factor in the variations of both OH and CO_2.

Models of these airglow emissions[1] suggest smooth layers with relatively little high spatial frequency content. This experiment is designed, in part, to verify these models.

Viewing the airglow emissions in the earth-limb mode provides a direct method of obtaining the vertical distribution provided the telescope has sufficient resolution. The earth-limb mode also provides an approximate gain of 60 for measurements of optically thin atmospheric emissions compared with vertical viewing in typical rocket-sonde probes.[5]

9.3 SPECIFICATIONS

The system performance objectives require three band measurements. The wavelength specifications for each are given in Table 9.1.

The requirement to map the structure of the airglow layers establishes the resolution requirement as given in Table 9.1. This is accomplished by employing three detector arrays with a telescope and a scanning mirror to

Table 9.1   ELIAS INSTRUMENTATION SPECIFICATIONS

1. Spectral domain:

Item	Species	Center wavelength, μm
1	N_2^+	0.3907
2	OH	2.7
3	CO_2	4.3

2. Spatial domain:

   Full field at the earth tangent: 9×9 km

   Resolution: 1.5 km

   Off-axis response to earth at 2°: less than NES

   Registration: In-track separation ≤ 0.5 pixel

   Cross-track position ≤ 0.5 pixel

3. Noise equivalent sterance (NES) [radiance]: Equal to or less than 3×10^{-7} W cm^{-2} sr^{-1} each wavelength
4. Signal-to-noise ratio (SNR): 100
5. Dynamic range: 10^5 with four linear channels each wavelength
6. Radiometric accuracy: $\pm 15\%$
7. Temporal domain: Scan rate, 10 per second
8. Rocket height: 250 km

---

map a 6×6 km region of the earth limb in a "stare" mode. Inactivation of the scanning mirror provides constant-height earth-limb scans as the payload moves forward in its trajectory.

The attitude control system (ACS) is used, under ground control, to direct the sensor optical axis at auroral forms as observed with an on-board bore-sighted TV camera.

Correlations among hydroxyl, carbon dioxide, and nitrogen emissions require spectral registration and geometric fidelity. Table 9.1 indicates band-to-band registration of ≤ 0.5 pixel for in-track and cross-track separation.

Radiometric accuracy is important in determining constituent number densities in the atmosphere and in modeling auroral enhancement mechanisms. The radiometric performance goal is stated in terms of a signal-to-noise ratio (SNR) that will provide an absolute uncertainty of $\pm 15\%$.

9.4   PERFORMANCE GOALS

Before a feasiblity study (trade-offs and iteration) can be undertaken, it is appropriate to summarize the specifications in terms of a list of primary system performance goals as follows:

1. Signal-to-noise ratio
2. Field of view (FOV)—high resolution with high off-axis rejection
3. Band-to-band registration
4. Radiometric accuracy
5. Practical and economical design

Any one of the objectives might be easily attained, but it is more difficult to achieve them simultaneously in one instrument. This is because these requirements have conflicting interdependent design implications. Perhaps the most seriously interdependent parameters are spectral resolution and SNR. High-spatial-resolution systems require large apertures, to maintain acceptable throughput, and high-D^* detectors. The optical design also has a strong influence upon the band-to-band registration and off-axis rejection.

9.5 RADIOMETRIC TRADE-OFF ANALYSIS

The radiometric analysis and trade-off study is illustrated in Fig. 9.2. The detector size and pitch result from a consideration of noise and cost. The required resolution, detector size and pitch, and relative aperture determine the focal length and aperture size. Finally, the dwell time and required SNR determine the detector D^* for a given size instrument. The feasibility of the design depends totally upon the availability of detectors with adequate detectivity.

9.5.1 Detector Considerations

Detector technology is the dominant driver for most electro-optical systems, so detector size is a good point at which to begin a feasibility study. Throughput increases with detector size, but larger detectors generally exhibit increased noise because of the area-related capacitance. Larger

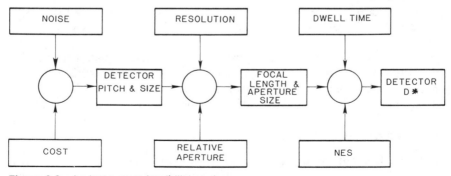

Figure 9.2 An instrument feasibility study.

detectors also generally result in larger and more costly systems; therefore, detectors of the order of 10^{-4}-cm^2 area (1-mm diameter) or smaller are desirable.

Detector pitch, optical focal length, aperture size, and dwell time are the basic parameters that determine the physical size and radiometric performance of the system. These numbers are derived from the optical throughput, modulation transfer function (MTF), and SNR analysis.

9.5.2 Resolution

The geometry of the rocket trajectory is given in Fig. 9.3. The earth depression angle α_e is the angular measure of the edge of the earth below the local horizontal. It is obtained from Fig. 9.3 by

$$\alpha_e = \arccos\left(\frac{r}{r + x}\right) \tag{9.1}$$

where $r = 6371$ km is the earth radius and x is the rocket altitude. The angle tangential to a shell at a height h (see Fig. 9.1) is given by

$$\alpha_h = \arccos\left(\frac{r + h}{r + x}\right) \tag{9.2}$$

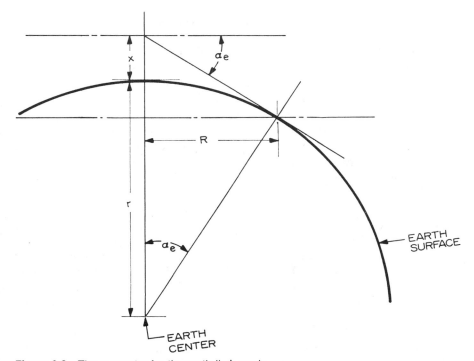

Figure 9.3 The geometry for the earth-limb angle.

**Table 9.2 DEPRESSION ANGLES TO THE EARTH EDGE α_e,
THE 87-km TANGENTIAL LAYER HEIGHT α_h, DIFFERENCE ANGLE $\Delta\alpha(R)$,
AND LAYER THICKNESS Δh VS. ROCKET HEIGHT FOR 1 mrad FOV**

Rocket height, km	α_e, deg	α_h, deg	$\Delta\alpha(R)$, deg	Δh, km
200	14.17	10.64	3.53	1.21
250	15.80	12.74	3.06	1.46
300	17.25	14.52	2.73	1.67
350	18.57	16.08	2.49	1.86

The rejection angle $\Delta\alpha(R)$ between the tangential height layer h and the earth yields a criterion for telescope rejection and is

$$\Delta\alpha(R) = \alpha_e - \alpha_n \qquad (9.3)$$

The thickness of a layer, Δh, at the tangential height, resolved by a $\Delta\alpha$ field of view is given by

$$\Delta h = (r + x)\sin\alpha_h \sin\Delta\alpha \qquad [\text{km}] \qquad (9.4)$$

The solutions to Eqs. (9.1) through (9.4) are given in Table 9.2 for a tangential height of 87 km (the peak of the OH layer) and a field of view of 1 mrad.

The data in Table 9.2 can be interpreted in terms of telescope rejection angles required, which vary from 2.5 to 3.5° depending upon the rocket height. The layer thickness varies between 1.2 and 1.86 km and is representative of the footprint achieved for a 1-mrad (0.057°) field of view.

The sampling theorem, as applied to the problem of measuring spatial structure, is that at least two samples per cycle must be obtained for the highest spatial frequencies. A practical implementation calls for five samples per cycle to avoid aliasing, since the MTF (Chap. 13) does not perform as an ideal spatial filter. The primary objective of this measurement is to map the structure of the airglow layers over a range of about 6 km.

A maximum spatial frequency of 1 cycle per 6 km is adequate to define the general layer shape and suggests a five-element detector array. The MTF generally exceeds 0.75 whenever the sample frequency is 5 times the sampled frequency. The detector configuration is illustrated in Fig. 9.4.

The desired spatial information is contained in the modulation components of the instrument output signal. The MTF analysis provides a measure of the loss in SNR for spatial frequencies of interest and must be included in the radiometric analysis.

9.5.3 Relative Aperture

The system throughput must be maximized in order to maximize the SNR. The throughput can be maximized, for fixed field of view, only by increas-

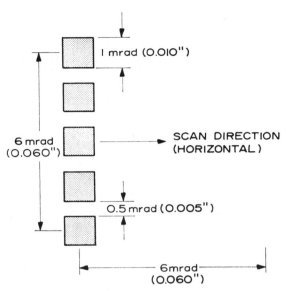

Figure 9.4   Layout of the ELIAS detector array.

ing the collector diameter (or, what is equivalent—by reducing the relative aperture or f-number). Cost and physical constraints suggest that an $F = 2$ system is reasonable.

A five-element linear detector array with pitch of 0.015 in (0.59 mm), square aperture of 0.010 in (0.39 mm), and an area of 6.45×10^{-4} cm^2 is available in InSb and Si. The detector throughput is therefore

$$\Upsilon = A_d \frac{\pi}{4F^2} = 6.45 \times 10^{-4} \frac{\pi}{4 \times 2^2} = 1.27 \times 10^{-4} \text{ cm}^2 \text{ sr} \quad (9.5)$$

The required FOV is 1×1 mrad or 10^{-6} sr. According to the invariance theorem, the entrance aperture throughput is

$$\Upsilon = A_c \Omega_c = A_c \times 10^{-6} = 1.27 \times 10^{-4} \text{ cm}^2 \text{ sr} \quad (9.6)$$

from which the aperture area is found to be

$$A_c = 1.27 \times 10^{-4}/10^{-6} = 1.27 \times 10^2 \text{ cm}^2 \quad (9.7)$$

and the effective aperture diameter is 12.7 cm (5 in.). The effective focal length is the product of the effective aperture and the relative aperture, or 25.4 cm (10 in.).

9.5.4   Sample Time

The purpose of this section is to determine the electrical subsystem bandwidth so that it will not limit the resolution obtained optically (see Sec. 7.2). The electrical bandwidth is also the noise bandwidth Δf, which is a figure of merit required to calculate the SNR. The specifications require a scan rate of 10 scans/s (Table 9.1), and each scan consists of six resolution elements of 1 mrad each. The dwell time T_d is defined as the time that the system dwells upon a single resolution element. The dwell time is the same as the integration time in this case. It is given by

$$T_d = \frac{1}{NS} = \frac{1}{60} = 1.67 \times 10^{-2} \text{ s} \qquad (9.8)$$

where N is the number of resolution elements per scan and S is the scan rate.

The resolution element can be visualized as an ideal square pulse. The objective is to design the electronics fast enough so the output voltage achieves the peak value, within 1%, before the pulse ends (Fig. 7.1). The system response V_t is given in terms of the pulse height $V(\text{max})$, the system time constant T_c, and the dwell time T_d as

$$\frac{V_t}{V(\text{max})} = 1 - \exp\left(-\frac{T_d}{T_c}\right) = 0.99 \qquad (9.9)$$

where $T_d = 5T_c$. The frequency response required is given by

$$f_2 = 1/2\pi T_c = 5/2\pi T_d = 47.7 \text{ Hz} \qquad (9.10)$$

The system performance can be significantly improved for most detectors by chopping the incoming flux so the detector output can be processed in ac amplifiers. The chopping frequency f_c must be 3 to 5 times the highest modulation frequency, 60 Hz, according to the sampling theorem. Thus, the chopping frequency must be 180 to 300 Hz, depending upon how fast the limiting electrical filters roll off.

9.5.5   Detector D^*

The detector D^* required to meet the SNR = 100 specification can be determined using Eqs. (5.28) and (6.3):

$$\Phi_{\text{eff}} = L_s(\lambda)\, \Delta\lambda\, \tau_e \tau_p A_{fs} \pi / 4\beta F^2 = 6.35 \times 10^{-12} \text{ W} \qquad (9.11)$$

where $L_s(\lambda)\, \Delta\lambda = 3.0 \times 10^{-7} \text{ W cm}^{-2} \text{ sr}^{-1}$

$$\tau_e = 0.5$$
$$\tau_p = 1$$
$$A_{fs} = A_d = 6.47 \times 10^{-4} \text{ cm}^2$$
$$F = 2$$
$$\beta = 3$$

The path losses are assumed to be zero, since the objective is to measure the total integrated sterance [radiance] along the path.

For SNR = 100, the noise equivalent power (NEP) must be

$$\text{NEP} = 6.35 \times 10^{-12}/(\sqrt{2} \times 100) = 4.49 \times 10^{-14} \text{ W}$$

Then the detectivity is

$$D^* = (A_d f_2)^{1/2}/\text{NEP} = 3.91 \times 10^{12} \text{ cm Hz}^{1/2}/\text{W} \qquad (9.12)$$

where $f_2 = \Delta f = 47.4$ Hz is the information-noise bandwidth and the factor $\sqrt{2}$ is introduced to account for the noise increase with coherent rectification (dc restoration). The only detectors that qualify at 4.3 μm are those that must be operated at reduced temperatures.

Indium antimonide (InSb) photovoltaic detectors,[6] which are available in linear arrays, are the most likely candidates for the 4.3-μm channel, since they respond to 5 μm. However, to obtain D^* values approaching 10^{12} requires "low-temperature background" operation (Chap. 16).

Low-background conditions are achieved by cooling the entire optical subassembly, including the telescope and focal plane (detector array). In this case, a liquid nitrogen dewar must be utilized as the basic housing for the optical section. This results in an operating temperature of approximately 80 K. Under these low-background conditions, the detector is not background-limited, because the photon noise is essentially zero. In this case, an optimum engineering design is one in which the detector thermal-resistance noise is the fundamental limitation.

The most significant parameter by which InSb detectors can be specified is the low-background $R_0 A$ product. This is the product of the detector resistance at 0 V and its area. A typical value for small-area detectors is 1×10^6 ohms cm^2. The detector resistance for the proposed size [Eq. (9.7)] is given by

$$R_0 = \frac{1 \times 10^6}{6.47 \times 10^{-4}} = 1.55 \times 10^9 \text{ ohms} \qquad (9.13)$$

and the limiting thermal noise current is

$$i_n = \left(\frac{4kT\Delta f}{R_0}\right)^{1/2} = \left(\frac{4 \times 1.38 \times 10^{-23} \times 80 \times 47.7}{1.55 \times 10^9}\right)^{1/2}$$
$$= 1.16 \times 10^{-14} \text{ A rms} \qquad (9.14)$$

The detector current responsivity is determined primarily by materials and is about 2.0 A/W at 4.3 μm. Thus, the NEP is

$$\text{NEP} = i_n/\mathcal{R}_c = 1.16 \times 10^{-14}/2.0 = 5.8 \times 10^{-15} \text{ W} \qquad (9.15)$$

The SNR is given by

$$\text{SNR} = \Phi_{\text{eff}}/\text{NEP} = 6.35 \times 10^{-12}/(\sqrt{2} \times 5.8 \times 10^{-15}) = 774 \quad (9.16)$$

(assuming an optical efficiency $\tau_e = 0.25$), which is better than the required value. See Table 9.1.

The current responsivity at 2.7 μm is down by the factor 0.67 (Fig. 8.1), which yields

$$\text{SNR} = 774 \times 0.67 = 519 \tag{9.17}$$

which is also acceptable.

The current responsivity at 0.4 μm is down by the factor 0.1, which yields

$$\text{SNR} = 774 \times 0.1 = 77 \tag{9.18}$$

which is not acceptable.

Silicon photovoltaic detectors with enhanced UV response are available in arrays.[7] The current responsivity at 300 K is given by the manufacturer as 0.18 A/W at 0.4 μm. They claim an improvement at 80 K of 22%, yielding $\mathcal{R} = 0.22$ A/W.

The detector noise also decreases with temperature so that the system becomes feedback-resistor thermal-noise-limited with cooling. The feedback-resistor noise voltage is given by

$$V_n = (4kTR\,\Delta f)^{1/2} = (4 \times 1.38 \times 10^{-23} \times 80 \times 10^{10} \times 47.7)^{1/2}$$
$$= 4.59 \times 10^{-5} \text{ V rms} \tag{9.19}$$

where $R = 1 \times 10^{10}$ ohms. This yields a low-temperature NEP of

$$\text{NEP} = \frac{4.59 \times 10^{-5}[\text{V}]}{0.22[\text{A/W}]1 \times 10^{10}[\text{ohms}]} = 2.09 \times 10^{-14} \text{ W}$$

at 0.4 μm.

Table 9.3 SUMMARY OF RADIOMETRIC TRADE-OFF ANALYSIS

Aperture diameter	12.7 cm (5 in.)
Focal length	25.4 cm (10 in.)
f-Number	2
IFOV at 250 km	1.46 km (1 mrad)
Throughput	1.27×10^{-4} cm^2 sr
Detector pitch	0.59 mm (0.015 in.)
Detector size	0.39×0.39 mm (0.010 in.)
Number of detectors	5 each band
Dwell time	1.67×10^{-2} s
Frequency response	47.7 Hz
Detectors	InSb (4.3 μm, 2.7 μm)
	Si (0.4 μm)
Optical efficiency	0.25
MTF	0.75
NESR	7.74×10^{-10} W cm^{-2} sr^{-1} (4.3 μm)
	1.66×10^{-9} (2.7 μm)
	2.80×10^{-9} (0.4 μm)
Off-axis angle to earth	3.06°

The SNR is given by

$$SNR = 6.35 \times 10^{-12}/(\sqrt{2} \times 2.09 \times 10^{-14}) = 215$$

which is adequate, provided the optical efficiency losses, including MTF, can be maintained at a factor of 0.50 or better.

Thus, the general system appears to be feasible with readily available detectors. Table 9.3 summarizes the radiometric trade-off analysis. The next step is to perform a configuration trade-off analysis.

9.6 OPTICAL CONFIGURATION TRADE-OFFS

The radiometric analysis above establishes the general parameters for adequate noise equivalent sterance [radiance] and sets the stage for optical configuration trade-offs. There are four general requirements for system performance with respect to the spatial domain that must be considered under configuration trade-offs:

1. Off-axis rejection to provide spatially pure measurements
2. Resolution required to map the spatial variations
3. Heat load on the cryogenically cooled baffles
4. Band-to-band registration

The first-order problem of earth-limb measurements is that of off-axis rejection or spatial purity. It is necessary to direct the optical axis of the telescope to within a few degrees of the earth. The earth is an extended-area source of infrared radiation (Fig. 9.5), which, in a poorly baffled system, would overwhelm the faint airglow emissions. Resolution requirements were outlined above.

The problem of heat load on the baffle relates to the requirements to minimize cryogen consumption. The total heat input to a 5-in. aperture from earth shine is only about 2 W and is the dominant cause of cryogen consumption. The length of the mission and the heat load determine the cryogen storage tank dimensions. In the case of ELIAS, the short (approximately 10-min) period of the flight makes this problem insignificant.

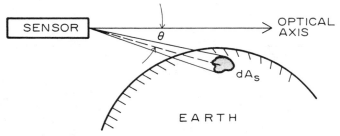

Figure 9.5   Illustration of the earth as an off-axis source.

The problem of heat load has been significantly reduced in some cases[8] by using specular baffles that retromap the aperture into itself. Off-axis rays that enter the aperture and strike the baffle surfaces are reflected so they pass out of the aperture and are not absorbed.

Band-to-band registration requirements are not difficult to achieve, but various configurations are considered below.

9.6.1 Off-Axis Rejection

Two generic configurations are possible whose principal axes are oriented perpendicular to the rocket trajectory. Figure 9.6 illustrates these alternatives.

The nth-order sunshade baffle (illustrated for $n = 2$) is designed to prevent direct illumination of the primary mirror (Chap. 14). This is accomplished by making the baffle angle α exclude the off-axis source so that radiation can reach the mirror only after scattering off the nth-order baffle. Off-axis energy enters the field of view only after successively scattering off the baffle and the primary mirror. Table 9.2 gives the off-axis range of the earth for the 87-km tangential layer at 3.06°. For a 5-in.

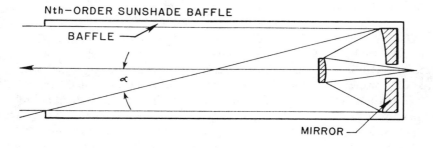

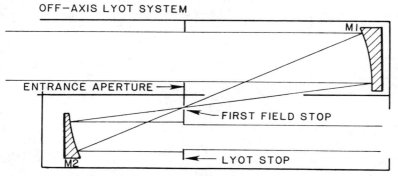

Figure 9.6 Candidate optical systems for off-axis baffling.

aperture, the baffle length is $5/\tan(3.06) = 93.5$ in. The sizing requirements do not permit a baffle that large.

The off-axis Lyot system is designed so the aperture stop is imaged at the Lyot stop (Chap. 14) to block diffracted energy. Since the mirror is illuminated directly by earth radiation, it must be given a low-scatter "superpolished" surface to minimize the scattering of off-axis energy into the field of view.[8] Such a system has been used successfully in the SPIRE program[5] and is the practical choice for ELIAS.

9.6.2 Band-to-Band Registration

There are four generic configurations possible to obtain band-to-band registration.[9] Each of these configurations is, in principle, compatible with the off-axis Lyot system. They are depicted in Fig. 9.7.

The first uses multiple optical systems, one for each band. Each individual system is relatively simple because only one band filter is required. However, the system suffers from a number of problems:

1. It requires duplication of the low-scatter Lyot system.
2. It has a parallax problem.
3. It suffers from cross-track misalignment when the focal lengths differ.
4. It is physically large.
5. The registration stability is poor because of the large tolerance buildup over large distances.

The second system avoids the problems of replicating the low-scatter optics and of cross-track misalignment due to differential focal lengths, as in the multiple optics system, and it is very compact. Registration stability is excellent because of mechanical compactness of the focal planes, but the bands are not registered along-track. Registering the images after recording can be accomplished only for systems that scan linearly along the track. This system is eliminated by the "stare"-mode requirements.

The third system, recently proposed by Jet Propulsion Laboratory,[10] is an imaging spectrometer. The objective forms an image on a slit that serves as the field stop and the entrance slit for a dispersion monochromator. In the focal plane, in one direction the location and size of the detector element determines its spectral bandpass, while in the other direction its size determines its cross-track footprint. All pixels are inherently registered in-track and cross-track. However, this system is eliminated, primarily because of the need for a new complex focal plane.

The fourth system, which was chosen for ELIAS, makes use of beam-splitting dichroic surfaces. Band-to-band registration can easily be achieved to the requirement of ± 0.5 pixel. Stability of registration is achieved in ELIAS as a by-product of the requirement to control the

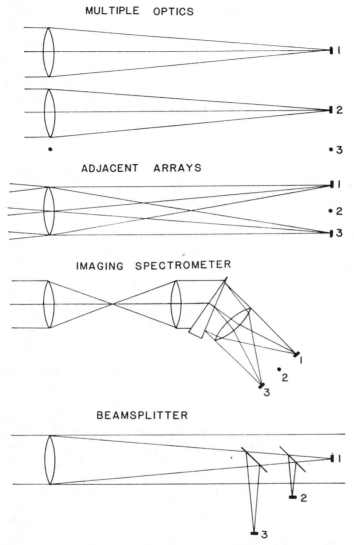

Figure 9.7 Candidate spectral separation and band-to-band registration systems.

temperature of the entire optical subsection in order to obtain enhanced detector performance by low-background operation. The beam-splitting subassembly shown in Fig. 9.7 is relatively compact, and registration can be accomplished at the subsystem level.

Figure 9.8 gives a pictorial representation of the ELIAS optical system, showing the LN_2 dewar and optical layout including telescope, baffles, scanning mirror, folding mirrors, dichroics, and detector-condensing mirrors.

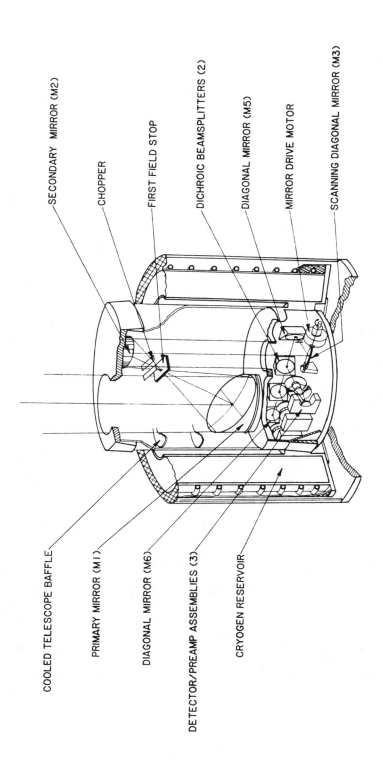

COOLED TELESCOPE BAFFLE

PRIMARY MIRROR (MI)

DIAGONAL MIRROR (M6)

DETECTOR/PREAMP ASSEMBLIES (3)

CRYOGEN RESERVOIR

SECONDARY MIRROR (M2)

CHOPPER

FIRST FIELD STOP

DICHROIC BEAMSPLITTERS (2)

DIAGONAL MIRROR (M5)

MIRROR DRIVE MOTOR

SCANNING DIAGONAL MIRROR (M3)

ELIAS SENSOR

Figure 9.8 Pictorial representation of ELIAS sensor illustrating LN$_2$ dewar, telescope, baffles, scanning mirrors, dichroics, and detector condensing optics.

9.7 RADIOMETRIC ACCURACY AND CALIBRATION TRADE-OFFS

Radiometric precision and accuracy[11] depend upon the detector operating conditions and calibration approach. The requirement for an optomechanical chopper depends upon the $1/f$ noise and drift and the required dynamic range. The low-background temperature required to achieve adequate detectivity provides for stable and repeatable operating temperatures. However, the large dynamic range of 1×10^5 precludes the use of dc amplifiers; thus, a chopper is necessary to avoid amplifier drift.

A fundamental limit to radiometric precision is set by system noise, which is characterized in terms of the noise equivalent sterance [radiance] and represents the spread in the output for a fixed input flux. The above radiometric analysis indicates that the specifications can be achieved.

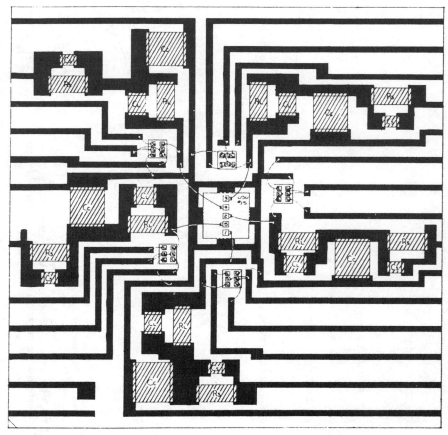

Figure 9.9 Five-element focal plane — detectors in center surrounded by chip resistors, FETs, and capacitors.

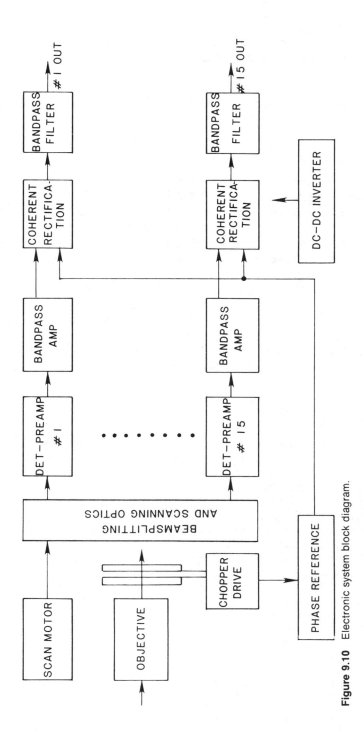

Figure 9.10 Electronic system block diagram.

The radiometric accuracy trade-offs include questions of laboratory calibration and inflight calibration techniques. The use of an independent high-temperature inflight calibration source is an option. However, the cold chopper provides a superior reference standard. Radiometric accuracy, therefore, depends primarily upon the uncertainty of the laboratory calibration standard source.

9.8 ELECTRONICS

The achievement of the specified precision requires that optimized individual electrometer preamplifiers be used with each of the detectors. Junction FETs exhibit lower noise than MOSFETs and function well at 80 K.

An important trade-off consideration relates to the focal-plane design. The requirement to reduce shunt capacitance and microphonic noise suggests a compact layout. The detectors, thick-film resistors, chip FETs, and capacitors can be bonded to a miniature printed circuit on a ceramic substrate, with thermally conducting epoxy for heat sinking, rather than using FETs in the usual TO-18 packages. This is illustrated in Fig. 9.9.

A low-power vibrating reed chopper[12] is located at the first field stop to produce an alternating flux at the detector. The chopper drive circuits, ac amplifiers, coherent rectifiers, and low-pass filters are of conventional design. The electronics system is represented as a functional flow block diagram in Fig. 9.10.

EXERCISES

1. A rocket instrumentation design requires a radiometer to measure the vertical distribution of atmospheric emissions. The sensor measures the overhead sterance as it moves upward through the layers at the rate of 10 km/s. It is required to resolve layers that are 1 km thick. The shape of the vertical distribution is the primary measurement objective. What should be the frequency response of the radiometer? What is the minimum chopping frequency? *Hint:* To measure shape accurately means to measure amplitude accurately; hence, set dwell time equal to 5 time constants [Eq. (7.2)].

2. Find the signal-to-noise ratio for a sequential scanning spectrometer with the following specifications:

> Collector diameter = 3 in.
> Collector field of view: 1 mrad (square) full-angle
> System f-number: 3
> Number of resolution elements: 50
> Scan rate: 2 scans/s
> Detector InSb:
>   Current responsivity: 2 A/W at 5 μm
>   Dark resistance: 1.9×10^9 ohms
> NES = 1×10^{-8} W cm^{-2} sr^{-1} at 5 μm
> $\beta = 3$; $\tau_e = 0.5$; $T = 300$ K

Hint: Use invariance theorem to find detector area. Find frequency response required based upon the sampling theorem. Solve for NEP.

REFERENCES

1. R. M. Nadile, A. T. Stair, Jr., C. L. Wyatt, and D. E. Morse, "Earth Limb Infrared Atmospheric Structure (ELIAS): A Remote Sensing Auroral and Airglow Shuttle Experiment," *Proc. Soc. Photo-Opt. Instrum. Eng.*, **265**, 304–309 (1981).

2. D. J. Baker, *Studies of Atmospheric Emissions*, *Final Report*, AFGL-TR-78-0251, Jan. 1978, p. 11.

3. K. D. Baker et al., "Measurement of 1.5 to 5.3 μm Infrared Enhancements Associated with a Bright Auroral Breakup," *J. Geophysical Res.*, **82**, 3525 (1977).

4. Ibid., p. 3518.

5. R. M. Nadile et al., "SPIRE—Spectral Infrared Experiment," *Proc. Soc. Photo-Opt. Instrum. Eng.*, **124**, 118–123 (1977).

6. Manufactured by Santa Barbara Research Center, Goleta, CA.

7. Manufactured by EG & G Electro-Optics, Salem, MA.

8. J. C. Bremer, "Baffle Design for Earth Radiation Rejection in the Cryogenic Limb-Scanning Interferometer/Radiometer," *Opt. Eng.*, **22**, 166 (1983).

9. A. M. Mika, "Design Trade-offs for a Multispectral Linear Array (MLA) Instrument," *Proc. Soc. Photo-Opt. Instrum. Eng.*, **345**, 23–31 (1982).

10. J. B. Wellman, "Technologies for the Multispectral Mapping of Earth Resources," *Proc. Int. Symp. on Remote Sensing of Environment*, **15**, 45–64 (1981).

11. C. L. Wyatt, *Radiometric Calibration: Theory and Methods*, Academic Press, N.Y., 1978, p. 5.

12. G. D. Frodsham, "Operation of Tuned Fork Light Choppers at Liquid Helium Temperatures," *Rev. Sci. Instrum.*, **46**, 312–316 (1975).

two

DETAILED DESIGN

Chapters 10 through 19 provide for *detailed design* of some configurations of some of the subsystems covered in Part 1. In particular, Chaps. 10 and 11 provide practical aspects of blackbody radiation and optical media that are useful in system design. This is accomplished by solving Planck's and Maxwell's equations. Chapters 12 through 18 provide design information for various sensor subsystems. Emphasis is placed upon achieving optimum design, considering the trade-offs that must be made to achieve system specifications. Chapter 19 discusses calibration and error analysis.

chapter *10*

Blackbody Radiation

10.1 INTRODUCTION

Sources were characterized in Chap. 3 in terms of *geometrical* radiant entities to provide useful and practical means to predict radiant energy transfer. Such a geometrical characterization of radiant sources is accomplished in terms of entities that serve as figures of merit in feasibility studies. The objective of this chapter is to provide detailed information on sources based upon blackbody radiation. A knowledge of these sources is useful in the choice of detectors, the design of optical systems, and in system evaluation and calibration.

In many applications, the source is not subject to control by the design engineer and originates as a result of physical processes in nature. Other sources used in electro-optical systems can be fabricated and are commercially available.

There exists a natural division of optical sources that depends upon the nature of the spectral distribution of the emitted energy. The first type of source radiates in a continuous fashion over a very broad band. A graph of its energy rate per unit wavelength is a smooth and continuous spectrum exhibiting a single maximum. This is typical of radiation resulting from the thermally excited molecular oscillation that is known as "blackbody radiation."[1-3]

The second form of optical source is that which radiates in a discontinuous fashion and is characterized by relatively strong emission in narrow spectral intervals but no radiation at all in other wavelength intervals. A graph of its energy rate per unit wavelength reveals a series of emission lines or bands. This is typical of radiation that results from changes in the electronic or molecular energy levels of atoms or molecules in a plasma. Figure 1.3 is an example of the emission band (showing individual lines) of the hydroxyl radical.

Electronic transitions result in relatively high energy changes, causing emission in the ultraviolet and visible regions. Vibration of the atoms that make up a molecule, or rotation of the molecule, results in relatively low energy transitions, causing emission in the infrared regions.

In this book, the major concern is with ideal *blackbody* radiation, which is useful in system design and calibration, rather than with theories that explain the physical processes giving rise to continuous or discontinuous radiation.

10.2 PLANCK'S EQUATION

All objects that have a temperature at any value other than absolute zero are continuously emitting and absorbing radiation. The radiation characteristics of ideal blackbody surfaces are completely specified if the temperature is known. Blackbody simulators are commercially available that constitute a good approximation to blackbody radiation over a useful wavelength region. The background in which target sources are embedded (see Fig. 2.3) or to which detectors are exposed can be conveniently modeled as blackbody radiation. Blackbody simulators are used as primary calibration standards. The accuracy of such a laboratory standard is primarily determined by the accuracy with which its temperature can be determined.

Blackbody radiation is described by *Planck's equation* where the value of the constants are as given in Appendix C. Equation (10.1) gives the spectral sterance [radiance] as a function of absolute temperature and wavelength.

$$L(\lambda) = \frac{2hc^2}{\lambda^5} \frac{1}{\exp(hc/\lambda kT) - 1} \qquad [\text{W m}^{-3}\text{ sr}^{-1}] \qquad (10.1)$$

where h is Planck's constant, 6.6262×10^{-34} Js
        c is the velocity of light, 2.9979×10^8 m/s
        λ is the wavelength in meters
        k is Boltzmann's constant, 1.3806×10^{-23} J/K
        T is absolute temperature in kelvins (K)

The unit for sterance [radiance] is given as watts per unit area per unit steradian per unit wavelength in Eq. (10.1), where the meter (m) is used as the unit of wavelength. The units m^{-3} should not be confused with volume concentration.

A convenient form of the equation that may be evaluated with a desk calculator is

$$L(\lambda) = \frac{1.191066 \times 10^4}{\lambda^5}$$
$$\times \frac{1}{\exp(1.43883 \times 10^4/\lambda T) - 1} \quad [\text{W cm}^{-2} \text{ sr}^{-1} \mu\text{m}^{-1}] \quad (10.2)$$

where λ is entered directly in micrometers (μm).

In many cases it is useful to solve Planck's equation for temperature given the sterance [radiance]. Equation (10.2) can be rewritten as follows:

$$T = \frac{1.43883 \times 10^4}{\lambda \ln[(1.191066 \times 10^4/L(\lambda)\lambda^5) + 1]} \quad (10.3)$$

The spectrum of an interferometer is more appropriately described in units of frequency or wavenumber. The wavenumber is

$$\bar{\nu}[\text{cm}^{-1}] = \frac{\nu[\text{Hz}]}{c[\text{cm/s}]} = \frac{1}{\lambda[\text{cm}]}$$

It is therefore appropriate to provide the spectral sterance [radiance] as a function of absolute temperature and wavenumber.

$$L(\bar{\nu}) = 2hc^2\bar{\nu}^3 \frac{1}{\exp(hc\bar{\nu}/kT) - 1} \quad [\text{W m}^{-1} \text{ sr}^{-1}] \quad (10.4)$$

The units for $L(\bar{\nu})$ are watts per unit area per unit steradian per unit reciprocal length, which is equivalent to W m^{-2} sr^{-1} m.

A convenient form of Eq. (10.4) is

$$L(\bar{\nu}) = \frac{1.19101 \times 10^{-12}\bar{\nu}^3}{\exp(1.4388\bar{\nu}/T) - 1} \quad [\text{W cm}^{-2} \text{ sr}^{-1} \text{cm}] \quad (10.5)$$

where $\bar{\nu}$ is entered directly in reciprocal centimeters, cm^{-1}.

Figures 10.1 and 10.2 illustrate the spectral distribution of blackbody radiation as a function of wavelength and wavenumber, respectively.

These curves illustrate the following:

1. The spectral radiant sterance [radiance] increases at all wavelengths for increased temperatures.
2. The peak of the curve shifts toward shorter wavelengths or longer wavenumbers for higher temperatures.
3. The ratio $\Delta L(\lambda)/\Delta T$ has its greatest value in those regions of the curve (Fig. 10.1) where the wavelength is less than the wavelength of the peak radiation, that is, the region to the left side of the peak. The ratio $\Delta L(\bar{\nu})/\Delta T$ has its greatest value in those regions of the curve (Fig. 10.2) where the wavenumber is greater than the wave-

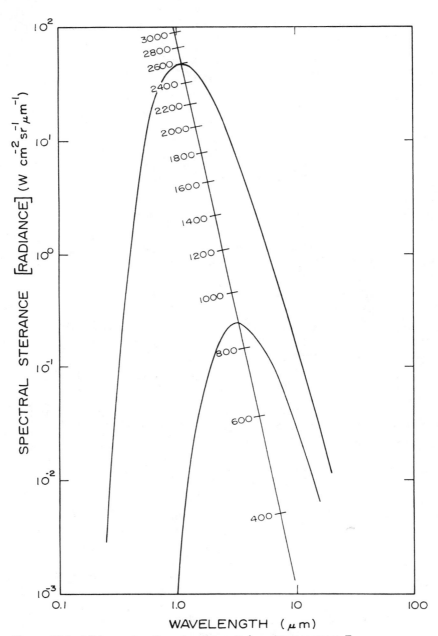

Figure 10.1 $L(\lambda)$ as a function of wavelength λ and temperature T.

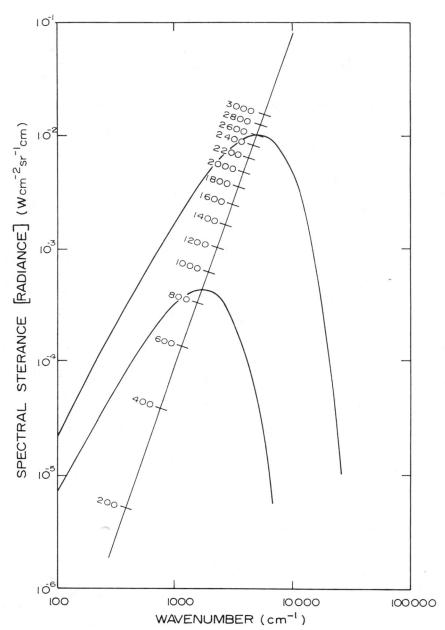

Figure 10.2 $L(\bar{\nu})$ as a function of wavenumber $\bar{\nu}$ and temperature T.

number of the peak radiation, that is, the region to the right side of
the peak.

4. The shape of the blackbody radiation curve is exactly the same for
   any temperature T.
5. A curve connecting the peak radiation for each temperature is a
   straight line.
6. The shape of the curve can be shifted along the straight line
   connecting the peaks to obtain the curve at any temperature.

A "do-it-yourself" slide rule can be constructed by placing a sheet of
tracing paper over Fig. 10.1 or Fig. 10.2 and tracing a curve and the line.
Then, by keeping the lines overlapping and setting the peak at the desired
temperature, the tracing becomes the blackbody curve for that temperature.

Tables have been published[4] that contain compilations of various
blackbody functions including the solution to Eq. (10.1) for spectral radiant
sterance [radiance]. The tables are useful for precision work; however,
interpolation is necessary for intermediate values. Slide rules are also
available that provide for rapid calculation of blackbody quantities with
good accuracy. However, the advent of the hand-held calculator has made
tables and slide rules obsolete.

The Planck radiation formula provides the basis for modeling numerous system design and analysis problems as illustrated in the following
examples.

Example 1: Find the temperature (in kelvins) of an extended-area
blackbody source that will produce a signal-to-noise ratio (SNR) of 10
at 22 μm.

Given: The sensor, an IR spectrometer, has a noise equivalent flux
(NEF) of 5×10^{-12} W cm^{-2} sr^{-1} μm^{-1}.

Basic equation:

$$T = \frac{1.43883 \times 10^4}{\lambda \ln\left[(1.91066 \times 10^4/L(\lambda)\lambda^5) + 1\right]} \quad (10.3)$$

Assumptions: (1) Sterance [radiance] is invariant. (2) The source is an
ideal blackbody.

Solution: An SNR of 10 and the invariance theorem require the source
sterance [radiance] to be 10 times the spectrometer NES or 5×10^{-11}
W cm^{-2} sr^{-1} μm^{-1}.

$$\ln\left(\frac{1.91066 \times 10^4}{5 \times 10^{-11} \times 22^5} + 1\right) = 18.122$$

$$T = \frac{1.43883 \times 10^4}{22 \times 18.122} = 36.09 \text{ K}$$

Thus, an ideal blackbody at a temperature of 36.09 K can be detected
with an SNR of 10 in this system.   ■

The parameters in Example 1 are typical of the state of the art of cryogenic sensor design. Arsenic-doped silicon (Si : As) semiconductor detectors operated at helium temperature and with low-temperature backgrounds are capable of measuring the heat emitted from extremely cold sources.

The following example follows from Example 1 and addresses the question of the temperature of the optics, which gives rise to the detector background in what are commonly referred to as "low-background" systems.

Example 2:   Find the temperature to which the optical subsystem must be cooled to eliminate the effects of self-emission upon the operation of the detector in a sensor design.

Given: The 1×1 mm cold detector is exposed to self-emissions of the optical subsystem through a 20° (half-angle) circular cold-shield aperture. The system throughput is 0.2 cm^2 sr; the optical efficiency, for external sources, is 0.4; and the noise equivalent sterance [radiance] is 5×10^{-12} W cm^{-2} sr^{-1} at 22 μm.

Basic equations:

$$\Phi = L\Upsilon \tag{3.21}$$

$$L(\lambda) = \frac{1.191066 \times 10^4}{\lambda^5} \frac{1}{\exp(1.43883 \times 10^4/\lambda T) - 1} \tag{10.2}$$

$$T = \frac{1.43883 \times 10^4}{\lambda \ln[(1.91066 \times 10^4/L(\lambda)\lambda^5) + 1]} \tag{10.3}$$

Assumptions: (1) A limiting solution is one for which the flux incident upon the detector from the baffle equals the detector NEP. (2) The self-emissions of the baffle and telescope are coded (chopped) and suffer the same losses as an external source. (3) The optical subsystem emits as a perfect blackbody.

Solution: The NEP on the detector is obtained using Eq. (3.21) and the optical efficiency for a noise equivalent sterance [radiance] of 5×10^{-12} W cm^{-2} sr^{-1}.

$$\text{NEP} = 5 \times 10^{-12}[\text{W cm}^{-2} \text{ sr}^{-1}] \times 0.2[\text{cm}^2 \text{ sr}] \times 0.4$$
$$= 4.0 \times 10^{-13} \text{ W}$$

The sterance [radiance] of the baffle is found that yields an NEP on the detector as a limiting case.

$$L = \frac{\text{NEP}}{\Upsilon \tau_e} = \frac{4.0 \times 10^{-13}}{3.67 \times 10^{-3} \times 0.4} = 2.72 \times 10^{-10} \text{ W cm}^{-2} \text{ sr}^{-1}$$

where the throughput for the optics is given by the detector area, the detector cold-shield solid angle, and the optical efficiency (0.4).

$$\Upsilon = 0.01(\text{cm}^2)\,\pi\,\sin^2 20° = 3.67 \times 10^{-3}\ \text{cm}^2\ \text{sr}$$

$$\ln\left(\frac{1.191066 \times 10^4}{2.72 \times 10^{-10} \times 22^5} + 1\right) = 15.955$$

$$T = \frac{1.43883 \times 10^4}{22 \times 15.955} = 40.99\ \text{K} \qquad \blacksquare$$

The use of the ideal assumption in Example 2 is extremely useful in system design because it represents a limiting case. The assumption that the baffles emit as a perfect radiator yields the worst-case solution. Such an assumption provides a relatively simple solution, since it is difficult to determine the real emissivity of mirrors, lenses, and baffles.

10.3   THE WIEN DISPLACEMENT LAW

The Planck radiation formula shows that the spectrum of the radiation shifts toward shorter wavelengths or longer wavenumbers as the temperature of the radiator is increased. The derivative of the Planck equation [Eq. (10.1)] with respect to wavelength yields the *Wien displacement law*, which gives the wavelength for which maximum radiation occurs for a given temperature[5]:

$$\lambda_m = 2898/T \qquad [\mu\text{m}] \qquad (10.6)$$

The solution to Eq. (10.6) provides for the wavelength designation along the straight line of Fig. 10.1.

The derivative of Eq. (10.4) with respect to wavenumber yields the wavenumber for which maximum radiation occurs for a given temperature:

$$\overline{\nu}_m = T/0.51 \qquad [\text{cm}^{-1}] \qquad (10.7)$$

The solution to Eq. (10.7) provides for the wavelength designation along the straight line of Fig. 10.2.

The wavelength λ_m, or wavenumber $\overline{\nu}_m$, at which the peak for maximum radiation occurs is significant for calibration purposes. At wavelengths less than λ_m or wavenumbers greater than $\overline{\nu}_m$, the spectral power density is changing very rapidly with temperature and wavelength or wavenumber, so a slight error in determining either the temperature or the wavelengths or wavenumbers results in a relatively large error in the calculated radiation (Chap. 19). However, at wavelengths beyond λ_m or wavenumbers less than $\overline{\nu}_m$, the radiation changes less rapidly and is therefore less sensitive to temperature errors.

Another factor that deserves attention is the uniformity (or lack of uniformity) of the radiation as a function of wavelength. It would be desirable to make use of a standard source with a uniform spectrum for the calibration of a spectrometer. This would result in uniform stimulation of the spectrometer response at all wavelengths throughout its free spectral range. However, the blackbody curve is very nonuniform except in the region of wavelengths near λ_m.

For these reasons, it is desirable to use relatively hot blackbody calibration sources. However, this calibration ideal is often difficult to follow because of the problem of operating and maintaining high-temperature blackbodies (that is, blackbodies that operate above 1000°C, where the materials begin to glow red hot and suffer oxidation). Also, such high-temperature blackbody sources tend to overdrive or saturate sensitive electro-optical sensors.

For example, the calibration of a UV photometer that is designed to measure the 3940-Å (0.3940-μm) molecular nitrogen first negative band[6] would, by Eq. (10.6), require a blackbody operated at a temperature greater than 7000 K to conform to the ideal outlined above. Tungsten lamps are often used as standard sources at temperatures as high as 2600 K.

10.4   THE STEFAN-BOLTZMANN LAW

The total power radiated per unit area of a blackbody is obtained by integrating Planck's radiation law over all wavelengths and is known as the *Stefan-Boltzmann law*[7]:

$$M = \varepsilon\sigma T^4 \quad \left[W/m^2 \right] \tag{10.8}$$

where ε may be called the "total hemispherical emissivity." For many purposes ε may be taken as a constant in the case of a solid and is characteristic of the solid *surface*. The *Stefan-Boltzmann constant* σ has a value of 5.66961×10^{-8} and has the units of W m^{-2} K^{-4}. At room temperature (approximately 300 K), a perfect blackbody ($\varepsilon = 1$) of area equal to 1 m^2 emits a total power of 460 W. If its surroundings are at the same temperature, it absorbs the same amount.

The heat loss from a blackbody at temperature T_1 to its surroundings at temperature T_2 is given by

$$M = \varepsilon\sigma\left(T_1^4 - T_2^4 \right) \quad \left[W/m^2 \right] \tag{10.9}$$

If the difference $\Delta T = T_1 - T_2$ is small, Eq. (10.9) can be written

$$M = 4\varepsilon\sigma T^3 \Delta T \quad \left[W/m^2 \right] \tag{10.10}$$

where T is the mean temperature $(T_1 + T_2)/2$.

The following example continues from Examples 1 and 2 and makes use of the *Stefan-Boltzmann law* to analyze the heat load on a cryogenic sensor.

Example 3:   Find the heat load on a satellite-borne cryogenic sensor resulting from earth radiations into the sensor baffle.

Given: The sensor is oriented along the zenith in a "lookdown" attitude at an orbital altitude of 250 km and has a 6-in. diameter baffle.

Basic equations:

$$\Phi = L\Upsilon \tag{3.21}$$

$$\alpha_e = \arccos\left(\frac{r}{r+x}\right) \tag{9.1}$$

$$M = \varepsilon\sigma T^4 \quad [\text{W/m}^2] \tag{10.8}$$

Assumptions: (1) The baffle is a perfect absorber. (2) The earth emits as an ideal blackbody at an average temperature of 247 K.

Solution: Equation (9.1) provides the angle, with respect to the local horizon, of the earth's tangent:

$$\alpha_e = \arccos\left(\frac{6371}{6371 + 250}\right) = 15.8°$$

The half-angle with respect to the local zenith is the complement of α_e, or 74.2°. Thus the aperture throughput, for the earth as a source is

$$\Upsilon = \frac{\pi}{4}(6[\text{in.}] \times 2.54[\text{cm/in.}])^2 \times \pi \sin^2(74.2) = 530.59 \text{ cm}^2 \text{ sr}$$

The earth sterance [radiance] is, by Eq. (10.8),

$$L_e = \frac{M}{\pi} = \frac{5.67 \times 10^{-8}[\text{W m}^{-2}\text{ K}^{-4}] \times 247^4[\text{K}^4] \times 10^{-4}[\text{m}^2/\text{cm}^2]}{\pi}$$
$$= 6.72 \times 10^{-3} \text{ W cm}^{-2}\text{ sr}^{-2}$$

where $\varepsilon = 1$.

The flux incident upon the aperture is, by Eq. (3.21),

$$\Phi = 6.72 \times 10^{-3}[\text{W cm}^{-2}\text{ sr}^{-1}] \times 530.59[\text{cm}^2\text{ sr}] = 3.5 \text{ W}$$

The above solution can be obtained more simply by the following: Assume the earth fills the sensor hemispheric field of view, i.e., $\Upsilon = A_c\pi$. Then

$$\Phi = L_e\pi A_c = M_e A_c = \sigma T^4 A_c$$
$$\Phi = 5.67 \times 10^{-8}[\text{W m}^{-2}\text{ K}^{-4}] \times 247^4[\text{K}^4] \times 10^{-4}[\text{m}^2/\text{cm}^2]$$
$$\times (6 \times 2.54)^2[\text{cm}^2] \times \pi/4 = 3.85 \text{ W}$$

Note: This result is very significant for liquid-helium-cooled sensors, since the heat of vaporization for liquid helium is approximately 1 W liter^{-1} h^{-1}. Thus, the sensor of this example requires 3.5 liters of liquid helium per hour to absorb the heat load of the earth emissions in a 6-in. diameter aperture. ∎

10.5 THE RAYLEIGH-JEANS LAW AND WIEN'S RADIATION LAW

Two well-known historical approximations[8] to Planck's law are readily obtained from Eq. (10.1). For the conditions $hc/\lambda kT \ll 1$, Eq. (10.1) reduces to

$$L(\lambda) \simeq 2ckT/\lambda^4 \quad [\text{W m}^{-3}\,\text{sr}^{-1}] \quad (10.11)$$

and is known as the *Rayleigh-Jeans law*, which is valid only at long wavelengths. For the conditions $hc/\lambda kT \gg 1$, Eq. (10.1) reduces to

$$L(\lambda) \simeq \frac{2hc^2}{\lambda^5} \exp\left(-\frac{hc}{\lambda kT}\right) \quad [\text{W m}^{-3}\,\text{sr}^{-1}] \quad (10.12)$$

and is known as *Wien's radiation law*, which is valid only at short wavelengths.

10.6 EMISSIVITY AND KIRCHHOFF'S LAW

As indicated above, practical sources approach the ideal blackbody. The ideal receiver and radiator of radiant energy is called a *blackbody radiator* for reasons given below.

If a small solid object S is located within an evacuated isothermal cavity, according to the second law of thermodynamics there will be a net flow of heat between the object and the walls toward the cooler of the two. Eventually, the object will come to equilibrium temperature with the cavity walls and will remain at that temperature. If the object absorbs only a portion, α, of the incident areance [exitance] E W/cm^2 that a perfectly black ideal radiator would emit at that temperature, then the object S will emit an amount εM equal to that absorbed; that is,

$$\alpha E = \varepsilon M$$

This is an expression of *Kirchhoff's law*, which states that the absorptivity α of a surface is exactly equal to the emissivity ε of that surface.[9]

The relationship

$$\Phi_i = \Phi_\alpha + \Phi_\rho + \Phi_\tau \quad [\text{W}] \quad (10.13)$$

(where Φ_i is the incident flux) is a statement of the conservation of energy. Dividing both sides by Φ_i yields

$$\alpha + \rho + \tau = 1 \quad [\text{unitless}] \quad (10.14)$$

For an opaque body, $\tau = 0$, so Eq. (10.14) becomes

$$\alpha = 1 - \rho \tag{10.15}$$

indicating that the surfaces of high reflectance are poor emitters. That is why the ideal emitter is literally a diffuse black surface, or a blackbody.

Generally the emissivity ε of a surface is a function of wavelength, temperature, and direction. For many cases of radiation in solids, ε can be considered constant. A radiating body is known as a *greybody* when $\varepsilon < 1$ and is independent of λ. The power spectral density curve of a greybody has the same shape as that for a blackbody, but at any wavelength it has a value that bears the ratio ε to that of an ideal blackbody.

EXERCISES

1. Verify Eq. (10.2) using Eq. (10.1) and the atomic constants in Appendix C. *Hint:* Use unit analysis to convert from meters to centimeters and micrometers.
2. Prove the Wien displacement law, Eq. (10.6), by taking the derivative of Planck's equation [Eq. (10.1)] with respect to wavelength.
3. Prove the Stefan-Boltzmann law, Eq. (10.8), by integrating Planck's equation [Eq. (10.1)] for all wavelengths.
4. Calculate the total radiative heat loss for an ambient temperature (300 K) blackbody of area equal to 1 m². *Note:* This corresponds roughly to the heat loss for a typical human body. Assume $\varepsilon = 1$.
5. Calculate the net radiative heat loss for a human body assuming a body temperature of 98°F when the walls, ceiling, and floor of a room are at a temperature of (a) 75°F; (b) 65°F. Assume body area is 1 m². *Note:* This problem illustrates the importance of the radiative temperature of the environment for human comfort.
6. Using Planck's equation, Eq. (10.1), written for wavelength, derive Eq. (10.4) for wavenumber, using the values for the atomic constants given in Appendix C and unit analysis.
7. An electronics box dissipates 10 W and has a total surface area of 120 in². Assume that the only mechanism for heat transfer out of the box is by radiation coupling into a background at 300 K. Find the equilibrium temperature for an emissivity of 0.1.
8. Calculate the spectral sterance [radiance] at 10 μm for a 300-K blackbody source in units of W cm^{-2} sr^{-2} μm^{-1}.
9. Calculate the spectral sterance [radiance] at 10 μm for a 300-K blackbody source in units of W cm^{-2} sr^{-1} cm.
10. Find the ideal blackbody temperature for calibrating a sensor at 10 μm. Assumption: Ideal temperature is one for which spectral sterance [radiance] is maximum at 10 μm.
    (a) Find T_{max} for $\lambda_{max} = 10$ μm.
    (b) Find T_{max} for $\nu_{max} = 1000$ cm^{-1}.
    Note: 1000 cm^{-1} = $(10^4 \mu m/cm)/10 \mu m$.
    (c) Explain why they differ.

11. Given that the energy in a photon is $\varepsilon = h\nu = hc/\lambda$, where ε = energy, h = Planck's constant, ν = optical frequency in Hz, and c = velocity of light, find the quanta rate for an energy rate of 10^{-12} W at 10 μm.

REFERENCES

1. J. C. DeVos, "Evaluation of the Quality of a Blackbody," *Physica*, **20**, 669 (1954).
2. W. L. Eisenman and A. J. Cussen, "A Comparative Study of Several Black Bodies," *Proc. IRIS*, **1**, 39 (1956).
3. C. S. Williams, "Discussion of the Theories of Cavity-Type Sources of Radiant Energy," *J. Opt. Soc. Amer.*, **51**, 564 (1961).
4. M. Pivovonsky and M. R. Nagel, *Tables of Blackbody Radiation Functions*, Macmillan, New York, 1961, 481 pp.
5. P. W. Kruse, L. D. McGlauchlin, and R. B. McQuistan, *Elements of Infrared Technology*, Wiley, New York, 1962, p. 29.
6. R. C. Whitten and I. G. Poppoff, *Fundamentals of Aeronomy*, Wiley, New York, 1971, p. 194.
7. P. W. Kruse, L. D. McGlauchlin, and R. B. McQuistan, *Elements of Infrared Technology*, Wiley, New York, 1962, p. 15.
8. R. A. Smith, F. E. Jones, and R. P. Chasmar, *The Detection and Measurement of Infra-Red Radiation*, Oxford University Press, London, 1957, pp. 25–35.
9. P. W. Kruse, L. D. McGlauchlin, and R. B. McQuistan, *Elements of Infrared Technology*, New York, 1962, p. 14.

chapter *11*

Optical Media

11.1 INTRODUCTION

An optical medium is described as part of the electro-optical system in Chap. 1 and is considered as such in Chap. 4 primarily in terms of flux transfer. The general subject of optical media includes atmospheric scattering, absorption, and emission, as well as the whole field of the interaction of electromagnetic radiation with materials. Thus it would not be possible to cover everything included under the heading "media" in any one book.

The objective of this chapter is to present some material of a general nature concerning the interaction of electromagnetic radiation with optical materials used in system design. This includes the effects of reflection, refraction, and polarization. In order to describe such effects it is necessary to specify the physical nature of the transverse wave motion associated with the propagation of electromagnetic energy.

It can be taken as fact[1] that the vibrations are perpendicular (transverse) to the direction of motion of the waves. This is based upon experimental evidence from polarization studies.

Maxwell's equations, written in differential form, provide the basis for the solution of the boundary problem associated with nonconductive dielectric media such as optical lenses and filters. They are

$$\text{curl } \mathbf{E} = -\mu \frac{\partial \mathbf{H}}{\partial t}$$

$$\text{curl } \mathbf{H} = \varepsilon \frac{\partial \mathbf{E}}{\partial t} \qquad (11.1)$$

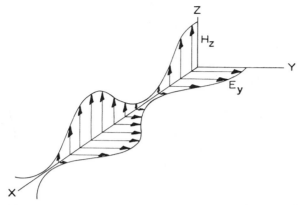

Figure 11.1 Illustration of the transverse-electric-magnetic (TEM) wave.

The simplest form of Maxwell's differential equations is for the transverse-electric-magnetic (TEM) mode, illustrated in Fig. 11.1, given by

$$\text{curl } \mathbf{E} = \frac{\partial E_y}{\partial x} = -\mu \frac{\partial H_z}{\partial t}$$

$$\text{curl } \mathbf{H} = \frac{\partial H_z}{\partial x} = \varepsilon \frac{\partial E_y}{\partial t} \tag{11.2}$$

which is obtained by expanding the curl of $\mathbf{E}$ and $\mathbf{H}$ [Eq. (11.1)] and writing the nonzero terms for a TEM wave.

Equations (11.2) can be solved (variables separated) by taking the derivative with respect to x and t and eliminating the common terms between equations to get

$$\frac{\partial^2 H_z}{\partial x^2} = -\mu\varepsilon \frac{\partial^2 H_z}{\partial t^2}$$

$$\frac{\partial^2 E_y}{\partial x^2} = -\mu\varepsilon \frac{\partial^2 E_y}{\partial t^2} \tag{11.3}$$

which are the wave equations in $\mathbf{E}$ and $\mathbf{H}$. The electromagnetic wave is a function of time, and the direction of propagation is given by the Poynting vector along the x axis (see Fig. 11.1):

$$\mathbf{P} = \mathbf{E} \times \mathbf{H} \tag{11.4}$$

where $\times$ signifies the cross product.

A useful form of the traveling wave equation for the electric vector is

$$E_y = E_{y0} \cos \omega \left(t - \frac{x}{v} \right) \tag{11.5}$$

as illustrated in Fig. 11.2. This wave can be shown to "travel" as follows: If $x = t = 0$, then $E_y = E_{y0}$ is the crest of the wave. This is also true for the case $t = x/v$, and since $x = tv$, the crest travels along x in time at a phase

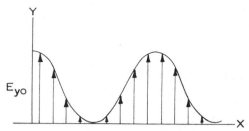

Figure 11.2   Illustration of a traveling wave.

velocity v. Equation (11.5) can also be expressed as

$$E_y = E_{y0} \exp\left[i\omega\left(t - \frac{x}{v}\right)\right] \tag{11.6}$$

The phase velocity can be obtained by taking the derivative of Eq. (11.5) twice with respect to x and then with respect to t and substituting the results into the **E**-field form of Eq. (11.3) to get

$$v = 1/\sqrt{\mu\varepsilon} \tag{11.7}$$

where $\varepsilon = \varepsilon_0 \varepsilon_r$ is termed the "permittivity" and $\mu = \mu_0 \mu_r$ is termed the "permeability," ε_r and μ_r are relative terms that are unity in free space, $\mu_0 = 4\pi \times 10^{-7}$, and $\varepsilon_0 = 1 \times 10^{-9}/36\pi$. The solution of Eq. (11.7) for free space yields the velocity of light, c.

The relative phase velocity is by definition

$$v_r = \frac{v}{c} = \left(\frac{\mu_0 \varepsilon_0}{\mu\varepsilon}\right)^{1/2} = \frac{1}{(\mu_r \varepsilon_r)^{1/2}} \tag{11.8}$$

The index of refraction for optical media is, by definition,

$$n = c/v = (\mu_r \varepsilon_r)^{1/2} = \varepsilon_r^{1/2} \tag{11.9}$$

since $\mu_r = 1$ for nonmagnetic substances.

The index of refraction n exhibits values for a changing field different from those it exhibits for a steady one.[2] Equation (11.9) yields the values for a steady field (which corresponds to zero frequency and consequently gives values that differ from the familiar numbers for visible light). The index of refraction is a function of frequency because of the nature of the interaction of the periodic force of the wave with the molecular structure of the material.

11.2   SNELL'S LAWS OF REFLECTION AND REFRACTION

Reflection, refraction, and polarization take place at the boundary between media that exhibit different properties as illustrated in Fig. 11.3. In the case of optical media the conductivity is usually zero.

The boundary conditions used are founded upon the principle of the conservation of energy stated as follows: The components of electric and

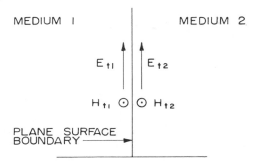

MEDIUM 1   MEDIUM 2

E_{t1} E_{t2}

H_{t1} ⊙ ⊙ H_{t2}

PLANE SURFACE
BOUNDARY

Figure 11.3 Boundary conditions for dielectric media.

magnetic fields parallel to the surface in one medium are at every point and every instant equal to the corresponding parallel components immediately on the other side of the boundary.[3] Thus there exists no discontinuity at the surface for parallel components of **E** and **H**.

Snell's laws of refraction and reflection are obtained using the above boundary conditions as illustrated in Fig. 11.4.

The phase velocity v of a wave front perpendicular to the direction of the ray is considered. The component of the wave along the $+x$ axis (tangential) is given by $x \sin \theta$, where θ is the angle of incidence. The component along the $-z$ axis (normal) is given by $-z \cos \theta$.

Thus, the exponential form of the traveling wave [Eq. (11.6)] can be written for each component:

$$\text{Incident:} \qquad \mathbf{E} = \mathbf{E}_0 \exp\left[i\omega\left(\frac{t - x \sin \theta - z \cos \theta}{v_1}\right)\right] \qquad (11.10a)$$

$$\text{Transmitted:} \qquad \mathbf{E} = \mathbf{E}_0 \exp\left[i\omega\left(\frac{t - x \sin \theta' - z \cos \theta'}{v_2}\right)\right] \qquad (11.10b)$$

$$\text{Reflected:} \qquad \mathbf{E} = \mathbf{E}_0 \exp\left[i\omega\left(\frac{t - x \sin \theta'' - z \cos \theta''}{v_1}\right)\right] \qquad (11.10c)$$

where θ, θ', and θ'' are the angles (with respect to the surface normal) of the incident, transmitted, and reflected components, respectively.

At the boundary, $z = 0$, all the exponents must be equal for the tangential **E** component; thus

$$\frac{\sin \theta}{v_1} = \frac{\sin \theta'}{v_2} = \frac{\sin \theta''}{v_1} \qquad (11.11)$$

The relationship between θ and θ' of Eq. (11.11) for the transmitted component is

$$\frac{\sin \theta}{v_1} = \frac{\sin \theta'}{v_2} \qquad (11.12)$$

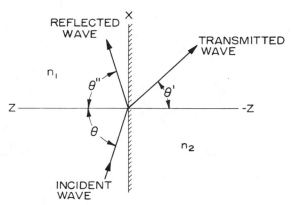

Figure 11.4 Geometry to illustrate Snell's law at the boundary of a dielectric material.

which, since $v = c/n$, is *Snell's law of refraction*

$$\frac{\sin \theta}{\sin \theta'} = \frac{n_2}{n_1} \tag{11.13}$$

The relationship between θ and θ'' of Eq. (11.11) for the reflected component is

$$\frac{\sin \theta}{v_1} = \frac{\sin \theta''}{v_1} \tag{11.14}$$

which is *Snell's law of reflection*; namely, that

$$\theta = \theta'' \tag{11.15}$$

Snell's law is utilized in lens design to trace rays through a system. Computer programs have been developed that yield the distribution of the energy in the focal plane from a point in the object plane. Such a ray trace gives the blur circle due to aberrations; however, it does not take into account diffraction effects.

Detailed lens design is beyond the scope of this book. However, Fig. 11.5 and the following example illustrate the use of Snell's law to calculate the lateral displacement of a ray in passing through a parallel-plate beam splitter. In actual application, such a device, referred to as a *dichroic*, reflects a designated band of wavelengths and transmits another band and functions as a color separation filter.

Example 1: Find the lateral displacement of a ray passing obliquely through a parallel-plate window.

Given: The angle of incidence with respect to the normal is 45°, the index of refraction of the window is 1.5, and the thickness is 4 mm.

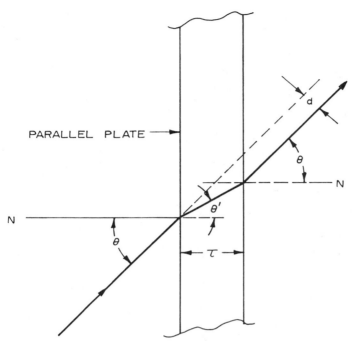

Figure 11.5 Illustration of the lateral displacement of a ray through a parallel-plate beam splitter.

Basic equation: Snell's law:

$$\frac{\sin \theta}{\sin \theta'} = \frac{n_2}{n_1} \tag{11.13}$$

Assumptions: The ray emerges from the second surface at 45° so that it continues in the same direction (see Exercise 2).

Solution: (Fig. 11.5)

$$\theta' = \arcsin\left(\frac{n_1}{n_2}\sin \theta\right) = \arcsin\left(\frac{1}{1.5}\sin 45°\right) = 28.13°$$

Path length s through the plate is

$$s = \tau/\cos \theta' = 4/\cos 28.13° = 4.54 \text{ mm}$$

The lateral displacement d is

$$d = s\sin(\theta - \theta') = 4.54\sin(45° - 28.13°) = 1.32 \text{ mm} \qquad ∎$$

11.3 FRESNEL'S FORMULA

In order to determine the magnitude of the electromagnetic fields of the transmitted and reflected waves it is necessary to take into account both the angle of incidence and the polarization of the reflected wave.

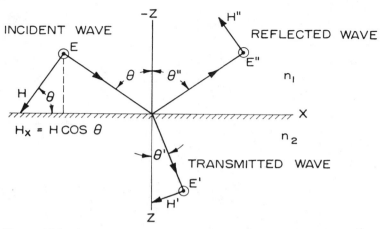

Figure 11.6 Geometry to illustrate perpendicular polarization at the surface of a dielectric material.

The boundary conditions are the same as those for Snell's law given above. In addition, the relationship between the **E** and **H** fields must be considered. The amplitude of the **E** field is greater in a nonmagnetic dielectric medium:

$$\mathbf{H} = \varepsilon_r^{1/2}\mathbf{E} = n\mathbf{E} \tag{11.16}$$

11.3.1   Perpendicular Polarization

The polarization of electromagnetic radiation has reference to the direction of vibration of the electric vector. Figure 11.6 illustrates perpendicular polarization. The electric vector is shown perpendicular to the xz plane, which is termed "the plane of incidence."

The boundary conditions are that the components of E_y and H_x that are tangent to the boundary between media of index n_1 and n_2 must be equal. (*Note:* The subscripts x and y are not used hereafter, and E and H are written in nonvector form.) In addition, in order to satisfy the principle of the conservation of energy, the sum of the reflected and transmitted components must equal the incident components. Thus

$$E - E'' = E' \tag{11.17}$$

and

$$(H + H'')\cos\theta = H'\cos\theta' \tag{11.18}$$

However, Eq. (11.18) can be written, using Eq. (11.16), as

$$n_1(E - E'')\cos\theta = n_2 E'\cos\theta' \tag{11.19}$$

Elimination of the transmitted wave E' between Eqs. (11.17) and (11.19) gives the Fresnel equation for reflection as

$$E - E'' = \frac{n_1}{n_2}(E + E'')\frac{\cos \theta}{\cos \theta'} \tag{11.20}$$

Substitute Snell's law, Eq. (11.13), and use

$$\tan \theta = \frac{\sin \theta}{\cos \theta} \tag{11.21}$$

to get

$$E - E'' = (E + E'')\frac{\tan \theta'}{\tan \theta} \tag{11.22}$$

Solving for the ratio of E'' to E yields the amplitude of the perpendicular reflection coefficient:

Reflection:
$$\frac{E''}{E_\perp} = \frac{\tan \theta - \tan \theta'}{\tan \theta + \tan \theta'} = \frac{\sin(\theta - \theta')}{\sin(\theta + \theta')} \tag{11.23}$$

The transmission amplitude is similarly obtained by eliminating E'' between Eqs. (11.17) and (11.19) to get

Transmission:
$$\frac{E'}{E_\perp} = \frac{2 \tan \theta'}{\tan \theta + \tan \theta'} = \frac{2 \sin \theta' \cos \theta}{\sin(\theta + \theta')} \tag{11.24}$$

11.3.2 Parallel Polarization

Parallel polarization is illustrated in Fig. 11.7, where the E vector is shown parallel to the xz plane of incidence. The boundary conditions are

$$(E + E'')\cos \theta = E' \cos \theta' \tag{11.25}$$

and

$$H - H'' = H' \tag{11.26}$$

Again, Eq. (11.16) is used to write Eq. (11.26) in terms of E:

$$n_1(E - E'') = n_2 E' \tag{11.27}$$

Elimination of the transmitted wave between Eqs. (11.25) and (11.27) yields the amplitude of the reflection coefficient

Reflection:
$$\frac{E''}{E_\parallel} = \frac{\tan(\theta - \theta')}{\tan(\theta + \theta')} \tag{11.28}$$

Similarly, the transmission amplitude coefficient is found:

Transmission:
$$\frac{E'}{E_\parallel} = \frac{2 \sin \theta' \cos \theta}{\sin(\theta + \theta')\cos(\theta + \theta')} \tag{11.29}$$

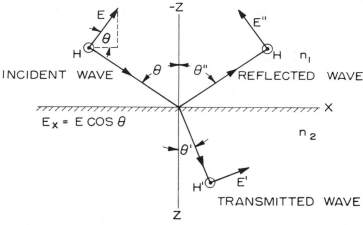

Figure 11.7 Geometry to illustrate parallel polarization at the surface of a dielectric material.

Flux is generally partially polarized. The flux exiting from a slit in a grating monochromator, or after reflectance from a folding mirror, for example, is polarized to some degree. The polarization sometimes results in metrology problems when the flux from various devices becomes cross-polarized. The following example illustrates the reflectance from an uncoated germanium beam splitter.

Example 2: Calculate the reflectivity of an uncoated germanium beam splitter for a perpendicularly polarized ray.

Given: The beam splitter is used at incident angle of 45°.

Basic equations:

$$\frac{\sin \theta}{\sin \theta'} = \frac{n_2}{n_1} \tag{11.13}$$

Reflection: $$\frac{E''}{E_\perp} = \frac{\tan \theta + \tan \theta'}{\tan \theta + \tan \theta'} = \frac{\sin(\theta - \theta')}{\sin(\theta + \theta')} \tag{11.23}$$

Assumptions: No absorptive losses.

Solution: The angle of refraction is given by Snell's law:

$$\theta' = \arcsin\left(\frac{n_1}{n_2}\sin \theta\right) = \arcsin\left(\frac{1}{4}\sin 45°\right) = 10.18°$$

The reflectivity is given by the square of the amplitude:

$$\rho = \frac{\sin^2(\theta - \theta')}{\sin^2(\theta + \theta')} = \frac{\sin^2(45° - 10.18°)}{\sin^2(45° + 10.18°)} = 0.48 \qquad \blacksquare$$

11.3.3 Normal Incidence

Normal incidence provides additional simplification. Consider the case where the quantity $\theta - \theta'$ approaches zero: Both the parallel and perpendicular reflectance coefficients reduce to

$$\frac{E''}{E} = \frac{\sin(\theta - \theta')}{\sin(\theta + \theta')} \tag{11.30}$$

since $\tan \theta = \sin \theta$ for small angles.

Equation (11.30) is written as

$$\frac{E''}{E} \cong \frac{n_2/n_1 - 1}{n_2/n_1 + 1} \tag{11.31}$$

where Snell's law [Eq. (11.13)] and the identity

$$\sin(A - B) = \sin A \cos B - \cos A \sin B \tag{11.32}$$

are used. Equation (11.31) becomes exact as $\theta - \theta' = 0$, for which

$$\frac{E''}{E} = \frac{n_2 - n_1}{n_2 + n_1} \tag{11.33}$$

which when squared yields the ratio of the *intensity* of the reflected and incident waves, or the reflectivity for normal incidence:

$$\rho_n = \left(\frac{n_2 - n_1}{n_2 + n_1}\right)^2 \tag{11.34}$$

Similarly, the amplitude of the transmission coefficients is obtained from Eqs. (11.24) and (11.29) as

$$\frac{E'}{E} \cong \frac{2n_1}{n_1 n_2} \tag{11.35}$$

which becomes exact as $\theta - \theta' = 0$. In order to obtain the transmissivity, a correction must be made to take into account the fact that the cross-sectional area of the transmitted beam is different from that of the incident beam. The transmissivity is the ratio of the energy of the transmitted beam to the total energy in the beam. The cross-sectional area of the beam in media 1 and 2 are proportional to the indices of refraction n_1 and n_2, and thus the transmissivity for normal incidence is

$$\tau_n = \frac{n_2}{n_1}\left(\frac{2n_1}{n_1 + n_2}\right)^2 = \frac{4n_1 n_2}{(n_1 + n_2)^2} \tag{11.36}$$

Equation (11.36) can be verified by again appealing to the principle of conservation of energy for nonabsorptive media, which in this case states that

$$1 = \rho + \tau \tag{11.37}$$

Equations (11.34) and (11.36) are substituted into Eq. (11.37) to get the identity

$$4n_1 n_2 + (n_2 - n_1)^2 \equiv (n_1 + n_2)^2 \tag{11.38}$$

Example 3: Find the normal reflectivity and transmissivity of an IR window manufactured from germanium.

Given: The index of refraction for germanium is 4.

Basic equations:

$$\rho_n = \left(\frac{n_2 - n_1}{n_2 + n_1}\right)^2 \tag{11.34}$$

$$\tau_n = \frac{n_2}{n_1}\left(\frac{2n_1}{n_1 + n_2}\right)^2 = \frac{4n_1 n_2}{(n_1 + n_2)^2} \tag{11.36}$$

Assumptions: There are no absorptive losses at the wavelengths of interest.

Solution:

$$\rho_n = \left(\frac{4 - 1}{4 + 1}\right)^2 = 0.36$$

$$\tau_n = \frac{4 \times 1 \times 4}{(1 + 4)^2} = 0.64$$

Note: The sum $\rho_n + \tau_n =$ unity.　　　　　　　　　　　　　　■

11.3.4　Polarizing Angle and Brewster's Law

For unpolarized light incident upon a dielectric medium such as glass as shown in Fig. 11.8, there are always a reflected ray and a refracted ray. Light is always polarized to some degree upon reflection from a surface. It was Brewster who discovered that light becomes completely plane-polarized at a particular angle.

　　This is illustrated in terms of the square of the amplitudes of the reflection coefficients (the reflectivity):

$$\rho_\perp = \frac{\sin^2(\theta - \theta')}{\sin^2(\theta + \theta')} \tag{11.39}$$

and

$$\rho_\parallel = \frac{\tan^2(\theta - \theta')}{\tan^2(\theta + \theta')} \tag{11.40}$$

for the special case where the reflected and refracted (transmitted) rays are

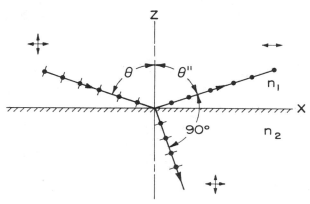

Figure 11.8 Illustration of Brewster's angle upon reflection from the surface of a dielectric medium.

exactly 90° apart, that is, when $\theta + \theta' = 90°$. In this case, where $\tan 90° = \infty$ and $\sin 90° = 1$ in Eqs. (11.39) and (11.40), the reflected light is totally perpendicularly polarized.

The special angle of incidence giving rise to the above conditions is called *Brewster's angle* or the *polarizing angle*. Since $\theta + \theta' = 90°$, $\sin \theta' = \cos \theta$ and Snell's law can be written as

$$\frac{\sin \theta}{\sin \theta'} = \frac{\sin \theta}{\cos \theta} = \tan \theta = \frac{n_2}{n_1} \qquad (11.41)$$

For the case where $n_1 = 1$ for air, Brewster's angle is

$$\theta_b = \arctan n_2 \qquad (11.42)$$

and the angle of incidence for total polarization depends only upon the index of refraction.

Reflection polarizers preferentially reflect and transmit radiation of orthogonal polarization. Numerous polarizers are available that capitalize on these principles.[4,5]

11.4 REFLECTANCE, TRANSMITTANCE, AND ABSORPTANCE

Equation (11.34) gives the reflectivity and Eq. (11.36) gives the transmissivity. Terms ending in -*ivity* refer to the ideal property of material having planar surfaces between two media and no oxides or coatings on the surface. Terms ending with -*ance* refer to the property of an actual sample or path.[6]

The actual reflectance and/or transmittance of a real object results from multiple reflection and transmission paths as illustrated in Fig. 11.9.

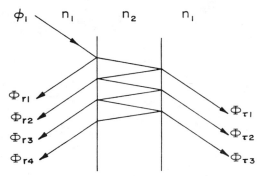

Figure 11.9 Illustration of the geometry for multiple reflection and transmission paths through a parallel plate.

The *reflectivity* is by definition the ratio of the reflected flux to the incident flux

$$\rho = \frac{\Phi_{r1}}{\Phi_i} \qquad (11.43)$$

and the *reflectance* is given by the ratio of the total reflected flux Φ_r to the incident flux

$$\rho_t = \frac{\Phi_r}{\Phi_i} \qquad (11.44)$$

However, the total reflected flux is given by

$$\Phi_r = \Phi_{r1} + \Phi_{r2} + \Phi_{r3} + \cdots \qquad (11.45)$$

which is

$$\Phi_r = \Phi_i \left\{ \rho + \rho(1 - \rho)^2 \exp(-2ad) + \rho^3(1 - \rho^2)\exp(-4ad) + \cdots \right\} \qquad (11.46)$$

where a is the absorption coefficient and d is the thickness. The reflectance ρ_t is

$$\rho_t = \frac{\Phi_r}{\Phi_i} = \rho + \frac{(1 - \rho)^2 \rho \exp(-2ad)}{1 - \rho^2 \exp(-2ad)} \qquad (11.47)$$

Similarly, the total flux transmitted is

$$\Phi_\tau = \Phi_{\tau1} + \Phi_{\tau2} + \Phi_{\tau3} + \cdots \qquad (11.48)$$

which is

$$\Phi_\tau = \Phi_i \left\{ (1 - \rho)^2 \exp(-ad) + 1 - \rho\right)^2 \rho^2 \exp(-3ad) \right\} \cdots \qquad (11.49)$$

and the transmittance is

$$\tau_t = \frac{\Phi_\tau}{\Phi_i} = \frac{(1 - \rho)^2 \exp(-ad)}{1 - \rho^2 \exp(-2ad)} \qquad (11.50)$$

The absorptance α_t is obtained between Eqs. (11.47) and (11.50) using the principle of the conservation of energy:

$$1 = \tau_t + \rho_t + \alpha_t \qquad (11.51)$$

again. In general, the absorptance is

$$\alpha_t = \frac{\Phi_\alpha}{\Phi_i} = \frac{(1 - \rho)[1 - \exp(-ad)]}{1 - \rho \exp(-ad)} \qquad (11.52)$$

Two cases are of interest:

1. For opaque material, the term $\exp(-ad) = 0$ and Eqs. (11.47) and (11.51) yield

$$\rho_t = \rho \quad \text{and} \quad \alpha_t = 1 - \rho_t \qquad (11.53)$$

Thus for opaque material ($\tau = 0$) the reflectivity is identical to the reflectance. It is also noted that surfaces of low reflectivity are good absorbers and appear to be black (in the visible).
2. For lossless media, $\exp(-ad) = 1$, for which Eqs. (11.47) and (11.51) yield

$$\rho_t = \rho + \frac{(1 - \rho)^2 \rho}{1 - \rho^2} \qquad (11.54)$$

and

$$\tau_t = \frac{(1 - \rho)^2}{1 - \rho^2} \qquad (11.55)$$

Equations (11.54) and (11.55) provide a good estimate of the reflectance and transmittance of flat optical windows in those spectral regions where the absorptance can be neglected.

The same equations hold when passing from high to low index of refraction. Total internal reflection will occur for all angles greater than the critical angle, which is given by Snell's law

$$\frac{\sin \theta}{\sin \theta'} = \frac{n_2}{n_1} \qquad (11.56)$$

where $n_1 > n_2$. The critical angle occurs for all $\theta' \geq 90°$ such that $\sin \theta' = 1$; thus,

$$\theta_c = \arcsin(n_2/n_1) \qquad (11.57)$$

EXERCISES

1. Calculate the relative phase velocity for germanium given that the index of refraction is 4.

2. Use Snell's law of refraction to prove that a ray incident upon the first surface of a parallel plate at an angle θ with respect to the normal emerges from the second surface at an angle equal to θ.

3. Calculate the reflectivity of glass ($n = 1.5$) as a function of incident angle between 0° and 90°. Plot the results for both perpendicular- and parallel-polarized light.

4. Find the critical angle for total internal reflection for glass ($n = 1.5$) and for germanium ($n = 4$).

5. Find the reflectivity and transmissivity for glass ($n = 1.5$), silicon ($n = 3.4$), and germanium ($n = 4$) for normal incidence. Assume no losses.

6. Find the reflectance and transmittance for glass ($n = 1.5$), silicon ($n = 3.4$), and germanium ($n = 4$) for normal incidence. Assume no losses.

7. Given an ray incident upon a surface of glass at an angle of 45°, find the refracted angle for $n_1 = 1$ and $n_2 = 1.5$.

8. Given that a fixed beam is reflected off a mirror in accordance with Snell's law, if the mirror is rotated by an increment of $\Delta\phi$, by what angle does the reflected beam increment?

REFERENCES

1. F. A. Jenkins and H. E. White, *Fundamentals of Optics*, McGraw-Hill, New York, 1957, p. 407.
2. Ibid., p. 482.
3. Ibid., p. 529.
4. M. Ruiz-Urbieta, E. M. Sparrow, and P. D. Parikh, "Two-Film Reflection Polarizers: Theory and Application," *Applied Optics*, **14**, 486–492 (1975).
5. D. T. Rampton and R. W. Grow, "Economical Infrared Polarizer Utilizing Interference Films of Polyethylene Kitchen Wrap," *Applied Optics*, **15**, 1034–1036 (1976).
6. P. W. Kruse, L. D. McGlauchlin, and R. B. McQuistan, *Elements of Infrared Technology*, Wiley, New York, 1962, pp. 138–142.

chapter *12*

Optical Systems

12.1 INTRODUCTION

The optical subsystem, as part of an electro-optical system, is characterized in Chap. 5 in terms of figures of merit that relate to its selective properties in the spatial, spectral, and polarization domains. This chapter is concerned with *system* design considerations rather than *component* design. The application of lenses and mirrors is emphasized.

The optical system design engineer is generally concerned with system performance, which requires a knowledge of elementary ray tracing and the configuration of telescopes. Once the layouts are completed, the detailed design is accomplished using computer techniques that are best consigned to the specialist and are beyond the scope of this book.

12.2 RAY TRACING

The ray or beam is defined in Sec. 3.2.4 in terms of the throughput of two finite areas. In this chapter we are concerned with the ray as a means of showing the *direction*[1] of the flow of flux, making use of the idea that radiant flux travels in straight lines (provided diffraction effects can be neglected).

Only the simplest form of ray tracing is required to create an optical system layout. This is accomplished by defining ideal lenses and mirrors for which perfect images are formed. Such a first-order design is of value to the optical designer to synthesize system configurations to achieve design goals. Once a system layout is accomplished, the effects of aberrations and diffraction can be considered.

The ideal lens is achieved, in the abstract, by the thin lens and paraxial ray assumptions. The ideal mirror is abstracted as one that has image-forming properties similar to those of the ideal thin lens. The thin lens and paraxial ray assumptions are defined below.

12.2.1   Conventions

Certain conventions and principles of optics are useful in ray tracing and designing optical layout drawings. The conventions are:

1. All figures are drawn with the rays traveling from left to right.
2. All distances along the axis are positive when measured from the associated vertex to the right and negative when measured to the left; this includes focal lengths and image and object distances.
3. All distances (object, image, or element heights) are positive when measured from the optical axis upward and negative when measured downward.

An important principle is the principle of *reversibility*, which states that if a reflected or refracted ray is reversed in direction, it will retrace its original path.[2] Thus, for example, a telescope and a collimator are visualized as systems that function in the reverse mode of one another.

12.2.2   Thin-Lens Ray Tracing

Thin-lens ray tracing is based upon the above conventions and the principle of reversibility plus an understanding of basic lens types as illustrated in Fig. 12.1. The lenses are referred to by the following terms: (*a*) double convex, (*b*) plano-convex, (*c*) positive meniscus, (*d*) double concave, (*e*) plano-concave, and (*f*) negative meniscus. It is noted that all positive lenses are thicker in the center than at the edge, while the opposite is true of negative lenses. Other distinctions useful in ray tracing are noted below in connection with the definitions for primary and secondary focal lengths.

A thin lens is defined as one whose thickness is considered small compared with distances associated with its optical properties such as focal length, radius of curvature, or diameter. In this case, refraction can be considered as occurring on a plane, normal to the optical axis, that coincides with the vertex as shown in Fig. 12.2.

The properties of an ideal image-forming thin lens are achieved by the assumption that the images correspond to those formed by paraxial rays[3]—rays for which the angles are small enough that the cosines equal unity and the sines equal the angles.

The above conventions are illustrated in Fig. 12.2. The lenses are converging, or positive, in parts (*a*) and (*c*) of the figure and diverging, or

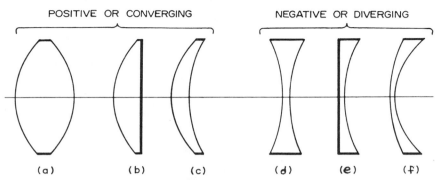

Figure 12.1 Illustration of cross sections of common types of lenses. (From F. A. Jenkins and H. E. White, *Fundamentals of Optics*, McGraw-Hill, New York, 1957, p. 28. Used with permission.)

negative, in parts (*b*) and (*d*). In addition, the *principal axis* is the line through the points F, F', and A, where the point F is the primary focal point, F' is the secondary focal point, and A is the lens vertex. The focal plane is a plane perpendicular to the axis that passes through the focal point.

The focal points are defined in reference to Fig. 12.2 as follows: The *primary* focal point F is a point on the optical axis such that any ray coming from it [Fig. 12.2(*c*)] or proceeding toward it [Fig. 12.2(*d*)] travels parallel to the axis after refraction. The distance f is the primary focal length. The primary focal point F is to the left of the vertex for a positive lens, and the focal length f is a negative number; F is to the right of the vertex for a negative lens, and f' is a positive number.

The *secondary* focal point F' is a point on the optical axis such that any ray moving parallel to the optical axis will, after refraction, proceed toward [Fig. 12.2(*a*)] or appear to come from [Fig. 12.2(*b*)] the secondary focal point. The secondary focal point F' is to the left of the vertex for a negative lens, and the secondary focal length f' is a negative number; F' is to the right for a positive surface, and f' is therefore a positive number.

There are two simple rules, derived from the definitions of the primary and secondary focal points, that can be used to accomplish a thin-lens ray trace. They are illustrated as follows:

1. The central ray, which passes through the vertex, is undeviated (see Fig. 12.3).
2. Parallel rays (oblique or on axis) converge toward or away from the point where the central ray intersects the focal plane after refraction.

The rules are illustrated for the case of image formation in Fig. 12.4. The central ray (1) is drawn first, undeviated from point P though the lens vertex (rule 1). A ray (2) is drawn parallel to the optical axis from point P

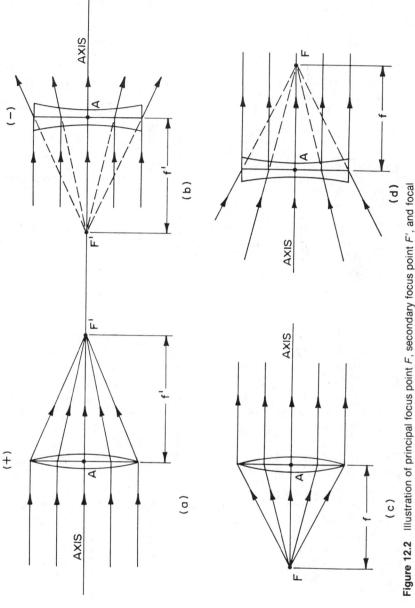

Figure 12.2 Illustration of principal focus point F, secondary focus point F', and focal lengths f and f', associated with negative ($-$) and positive ($+$) thin lenses. (From F. A. Jenkins and H. E. White, *Fundamentals of Optics*, McGraw-Hill, New York, 1957, p. 29. Used with permission.)

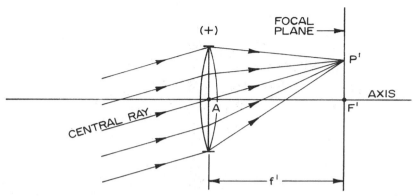

Figure 12.3 Illustration of how parallel rays are brought into focus at the point where the central ray intersects the focal plane.

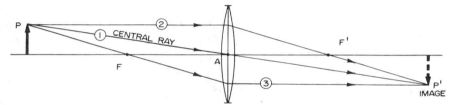

Figure 12.4 Illustration of parallel-ray method (rule 3) to locate the image formed by a thin lens. The numbers give the order in which the lines are drawn.

and passing through the focal point F' after refraction (rule 2). These two rays are sufficient to define the point P'. However, another parallel ray (3), drawn through F (using the principle of reversibility) after refraction, confirms the location.

The principle of reversibility can be illustrated in the case of Fig. 12.4. If the object is placed at the position previously occupied by its image, it will be imaged at the position previously occupied by the object. The object and image are interchangeable, or conjugate.[4] Any pair of object and image points such as P and P' are conjugate, and the planes through these points, perpendicular to the optical axis, are conjugate.

Figure 12.5 illustrates the method of tracing an arbitrary oblique ray (1) through the lens. A construction-line central ray (2) is drawn parallel to the arbitrary ray (rule 1). Then, ray (3) is drawn to intersect the central ray (2), after refraction, in the focal plane (rule 2).

12.2.3 Images

A real image can be made visible on a screen placed at the image point of Fig. 12.4. A virtual and erect image is formed if the object is placed inside the focus, as in the simple magnifier illustrated in Fig. 12.6.

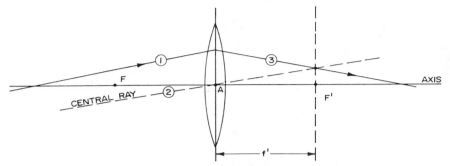

Figure 12.5   Illustration of method for graphically tracing an arbitrary oblique ray using the central ray (rule 1 and 2). The numbers correspond to the order in which the rays are drawn.

The magnifier (Fig. 12.6) is ray-traced as follows: The central ray (1) is drawn undeviated through P and A (rule 1). A ray (2) parallel to the axis is drawn through P, which after refraction (3) passes through F' (rule 2). An oblique ray (4) is drawn through F and P and emerges (5) parallel to the axis (rule 2). These rays (1, 3, or 5) do not converge to form a real image. However, they appear to be coming from the point P' on the far side of the lens; here, the virtual image is erect and magnified.

12.2.4   Lens Formula

The "lens formula" is useful in predicting the focal length required for giving image and object distances.[5] The formula is derived in reference to Fig. 12.7, where o and i are the object and image distances, respectively, and y and y' are the object and image heights, respectively.

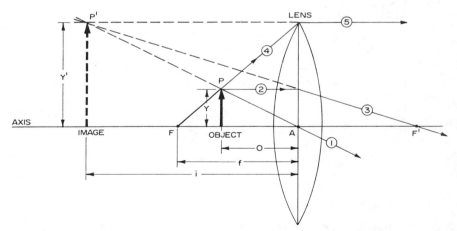

Figure 12.6   Illustration of a ray trace to locate the virtual image formed by a positive lens when the object is between the vertex A and the focus F. The numbers give the order in which the rays are drawn.

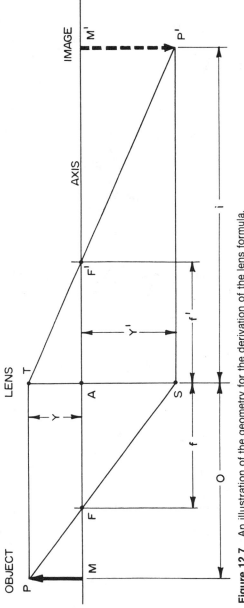

Figure 12.7 An illustration of the geometry for the derivation of the lens formula.

From similar triangles $TP'S$ and $TF'A$ we can write

$$\frac{y + y'}{i} = \frac{y}{f'} \tag{12.1}$$

From similar triangles PTS and FAS we can write

$$\frac{y + y'}{o} = \frac{y'}{f} \tag{12.2}$$

Taking the difference between these two equations yields

$$\frac{y + y'}{i} - \frac{y + y'}{o} = \frac{y}{f'} - \frac{y'}{f} = -\frac{y + y'}{f} \tag{12.3}$$

since $f' = -f$ by convention. The term $y + y'$ cancels from each term, yielding the lens formula

$$\frac{1}{i} + \frac{1}{f} = \frac{1}{o} \tag{12.4}$$

The lateral magnification is given by the ratio of y' to y, which by similar triangles TSP' and TSP is

$$m = \frac{y'}{y} = \frac{i}{o} \tag{12.5}$$

Example 1:   Given that for Fig. 12.7 the focal length is 6 units and the object distance is 9 units, find the image distance and magnification.

Basic equations:

$$\frac{1}{i} + \frac{1}{f} = \frac{1}{o} \tag{12.4}$$

$$m = \frac{y'}{y} = \frac{i}{o} \tag{12.5}$$

Assumptions: Thin lens and paraxial rays

Solution: By convention, $o = -9$ and $f = -6$, and thus from Eq. (12.4),

$$i = \frac{1}{1/o - 1/f} = \frac{1}{-1/9 + 1/6} = 18 \text{ units}$$

The positive 18 units indicates that the image is 18 units to the right of A.

The magnification is given by

$$m = \frac{i}{o} = \frac{18}{-9} = -2$$

The negative sign on the magnification indicates that P' is measured below the axis and that the image is inverted.   ■

The following example illustrates the solution of the lens formula for the magnifier of Fig. 12.6.

Example 2:   Given that for Fig. 12.6 the focal length F is 10 units, the distance to the object, o, is 5 units, and the height of the object is 3 units, find the location and height of the virtual image.

Basic equations:

$$\frac{1}{i} + \frac{1}{f} = \frac{1}{o}$$ (12.4)

$$m = \frac{y'}{y} = \frac{i}{o}$$ (12.5)

Assumptions: Thin lens and paraxial rays

Solution: By convention, $o = -5$, $y = 3$, and $f = -10$, and thus from Eq. (12.4),

$$i = \frac{1}{1/o - 1/f} = \frac{1}{-1/5 + 1/10} = -10 \text{ units}$$

The negative 10 units indicates that the image is 10 units to the left. The magnification is, by Eq. (12.5),

$$m = \frac{i}{o} = \frac{-10}{-5} = 2$$

The positive sign on the magnification indicates that P' is above the axis and that the image is erect. The height of the image, y', is, by Eq. (12.5),

$$y' = ym = 3 \times 2 = 6 \text{ units} \qquad ■$$

The "lensmakers' formula," which is given in the literature,[6] provides solutions for the radius of curvature for the first and second surfaces to achieve a given focal length as a function of the index of refraction of the material.

Lenses can be manufactured of various materials that transmit radiant energy at the wavelength of interest. The preferred shape is one that results in the least spherical aberration, which depends upon the material to be used. The shape factors for minimum blur are given in the literature[7] as a function of the index of refraction for various materials.

12.2.5 Mirror Ray Tracing

A reflective surface can be shaped to produce images similar to those of the ideal thin lens. As with the thin lens, we are interested in a simple form of

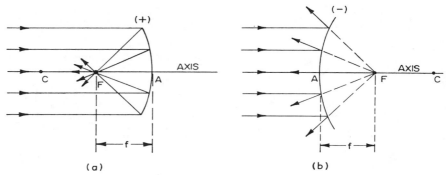

Figure 12.8 Illustration of focus point F associated with (a) positive (+) and (b) negative (−) mirrors.

ray tracing for which we define the mirror surface to be shaped to provide near-perfect images.

It is generally possible to find an optical layout using mirrors that is equivalent to that of a thin-lens system. However, there are some distinct advantages to the reflective systems: (1) The focal length is independent of wavelength; chromatic effects, in general, are absent, (2) Visual alignment techniques are possible, since the system passes visible light. On the other hand, mirrors do not offer the same possibilities for correction of spherical aberrations as do lenses; for this reason, camera lenses are generally refractive.

The conventions for ray tracing for mirrors are the same as for thin lenses with several exceptions. A mirror has only one focal plane, and the reflected ray travels from right to left. The mirrors are converging, or positive, in Fig. 12.8(*a*) and diverging, or negative, in Fig. 12.8(*b*),

The focal points are defined in reference to Fig. 12.8 as follows: the mirror exhibits a focal length that is defined such that any ray moving parallel to the optical axis will, after reflection, proceed toward [Fig. 12.8(*a*)] or appear to come from [Fig. 12.8(*b*)] the focal point. For a positive mirror, the focal point is to the left of the vertex and the focal length *f* is a negative number; for a negative mirror, *F* is to the right of the vertex and *f* is a positive number.

Ray tracing a reflective system is simpler than for a refractive system. Ideal mirror performance is achieved by the paraxial ray assumption, where all the rays pass through the focal point as illustrated in Fig. 12.8.

The rules for ray tracing a mirror are similar to those for a thin lens. It is convenient to define a ray that passes through the focal point *F* as the central ray. The rules are as follows:

1. The central ray (passing through *F*) travels parallel to the optical axis after reflection.
2. Parallel rays (oblique or on axis) converge toward or away from the point where the central ray intersects the focal plane after reflection.

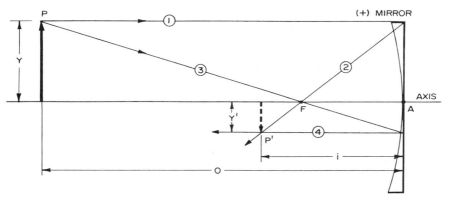

Figure 12.9 Illustration of parallel-ray method (rule 3) to locate the image formed by a converging mirror. The numbers give the order in which the lines are drawn.

Those rules are illustrated in the case of image formation in Fig. 12.9. A ray (1) parallel to the optical axis is drawn from the object point P to the mirror surface and through the focal point F (2) after reflection (rule 2). A central ray (3) is drawn from the point P through the focal point F and is drawn parallel (4) to the axis after reflection (rule 1). The intersection of rays 2 and 4 locate the image point P'. The points P and P' are conjugate points in object and image space.

Figure 12.10 illustrates the method to trace an arbitrary oblique ray (1). A central ray (2) parallel to the oblique ray (construction line) is drawn through the focal point F and is then drawn parallel to the axis (3) after reflection (rule 1). Ray 4, the continuation of ray 1, converges with ray 3 at P in the focal plane (rule 2).

The thin-lens formula, Eq. (12.4), can be applied to the spherical mirror; however, one change in the conventions must be made. Incident rays, after reflection, travel from right to left; the image distance measured from the vertex to the left must be considered a positive number.

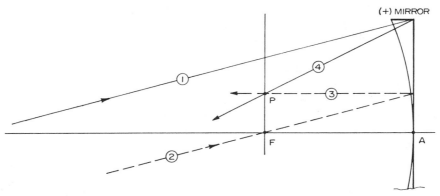

Figure 12.10 Illustration of the method to graphically ray trace an arbitrary oblique ray for a mirror (rules 1 and 2). The numbers give the order in which the rays are drawn.

Example 3:   Given that for the mirror of Fig. 12.9 the focal length is 10 units and the object distance is 25 units; find the image distance and the magnification.

Basic equations:

$$\frac{1}{i} + \frac{1}{f} = \frac{1}{o} \tag{12.4}$$

$$m = \frac{y'}{y} = \frac{i}{o} \tag{12.5}$$

Assumptions: Paraxial rays

Solution: By convention, $o = -25$ and $f = -10$. Thus from Eq. (12.4),

$$i = \frac{1}{1/o - 1/f} = \frac{1}{-1/25 + 1/10} = +16.67$$

The magnification is given by

$$m = \frac{i}{o} = \frac{+16.67}{-25} = -0.67$$

The positive sign on the image distance indicates that the image is to the left of the vertex, in accordance with the convention for mirrors, and the negative sign on the magnification indicates that the image is inverted.   ■

12.3   APERTURES AND STOPS

In Chap. 5 the optical subsystem is characterized in terms of the figures of merit—throughput and field of view. The effect of stops and apertures must be considered in order to define these figures of merit for a system.

In Fig. 12.11, a single-lens system is illustrated with two stops. The rays shown correspond to parallel rays from two points on a distant object. The size of the bundle of parallel rays is limited by the stop that is located on the lens. This stop is termed an *aperture stop* because it determines the amount of flux reaching the image from any point on the object. The second stop, placed in the focal plane, determines the extent of the object that can be represented in the image. This stop is termed the *field stop* because it determines the field of view.

The simple system illustrated in Fig. 12.11 is an extremely important one. The field stop is placed in the focal plane so that for objects at large distances the image is formed in the field stop and the field of view is very sharply defined.

The ideal sensor field of view is defined in Sec. 5.2 as one that has unity relative response over a specified region and zero response elsewhere. The cross section of such an ideal field of view can be defined as the

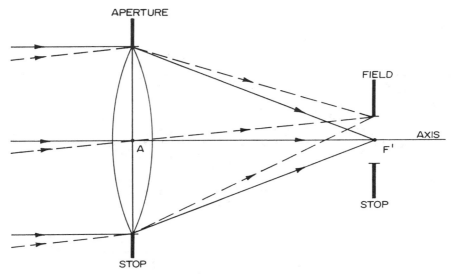

Figure 12.11 Illustration of aperture stop and field stop.

normalized angular response of the system to a point source (see Fig. 5.1). Since the image of a point source in the focal plane approaches a point (neglecting diffraction and distortion), the convolution of the point image with the field stop results in a field-of-view function that approaches the field-stop function.

12.3.1 Entrance and Exit Pupils

The edge of the lens always functions as an aperture stop; however, it may not be the limiting stop. It is necessary to consider the effect of stops that are not necessarily located in the focal plane or on the lens edge.

A stop $P'E'$, placed behind the lens as in Fig. 12.12, is in the image space and limits the image rays. *Image space* is defined as everything that pertains to the rays after refraction, and everything that pertains to them before refraction is defined as *object space*. For every postion on the object there is a corresponding position for the image. Thus, in this case, the image of the stop is in object space.

Figure 12.12 illustrates a case that behaves like a magnifier (Fig. 12.6) with the stop inside F'. Rays 1 through 4, using the parallel rule, locate the virtual image of the stop. As viewed from F, the stop in object space appears magnified at P, functions as the limiting aperture (as compared to the edge of the lens), and is called the *entrance pupil*. On the other hand, viewed from F', or beyond the stop in image space, this stop is the limiting aperture and is called the *exit pupil*.

Which stop functions as the limiting aperture depends upon the point from which it is viewed. For example, the edge of the lens in Fig. 12.12

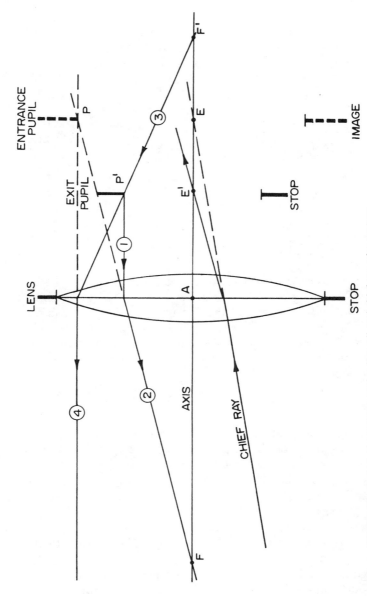

Figure 12.12 Illustration of entrance and exit pupil and chief ray of a system.

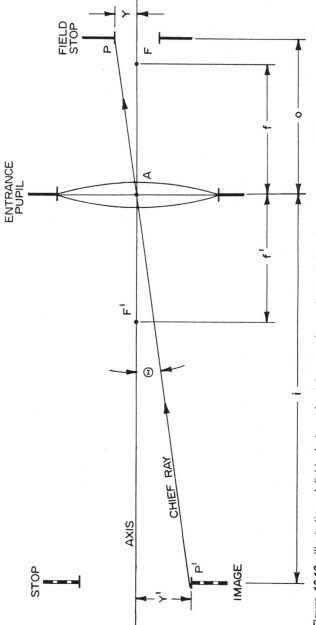

Figure 12.13 Illustration of field of view of a telescope focused at a finite object distance.

functions as the limiting aperture, or exit pupil, when viewed from within F' near the point E'. The general rule is that the limiting system aperture is the stop, or image of a stop, that subtends the smallest angle as seen from the *object* point.

In general, a limiting aperture is called an exit pupil when it is in image space and an entrance pupil when it is in object space. Any stop can be treated as an object, and its image is found using normal ray-tracing techniques as shown above. When the edge of a lens is considered as a stop, as in Fig. 12.11, the image and object of the stop coincide; hence, the lens edge functions as both entrance and exit pupil. Any ray in object space that passes through the center of the entrance pupil is termed a *chief ray*. Such a ray passes through, or appears to pass through, the center of the exit pupil after refraction. This follows, since for every point on the image there is a conjugate point on the object.

The use of the chief ray to determine the field of view is illustrated in Fig. 12.13, where the usual notation is reversed, insofar as primes and object and image distances are concerned, since the field stop is an object and we wish to find its image.

The field stop is located beyond the focal plane and corresponds to a telescope focused at a finite object distance. The image of the field stop, in the telescope's object space, determines its field of view. As viewed from the object point P', the system aperture is the lens edge, which is therefore both system entrance and exit aperture.

The *limiting* chief ray is directed from the edge of the object to the center of the entrance pupil. The half-angle field of view Θ is, in general, the smallest angle subtended at the center of the entrance pupil by any stop, or image of a stop, in object space.

By similar triangles we can write

$$i/y' = y/o \tag{12.6}$$

which is valid for $i = \infty$, which occurs when the field stop is moved to F. Thus, the half-angle Θ is always given by

$$\Theta = \arctan(y'/i) = \arctan(y/o) \tag{12.7}$$

In the special case where the system is focused at infinity and the field stop is at F, the field of view is

$$\Theta = \arctan(y/f) \tag{12.8}$$

12.4   ABERRATIONS

The image of a point source is not a point in practical optical systems. The optical blur is a nonzero area containing most of the flux incident upon the focal plane. The optical blur size is determined either by aberrations or by diffraction limits for corrected systems.

In principle, spherical aberrations can be avoided by making use of systems that are corrected (aspherical) through the use of elliptical or parabolic surfaces. Chromatic aberrations can be avoided by using reflective systems. Chapter 13 considers diffraction limits.

Reducing the optical system *f*-number until the optical blur is equal in area to the field stop provides a criterion for systems limited by spherical aberrations. This maximizes throughput at the sacrifice of the field-of-view function, which departs from an ideal spatial filter.

Therefore, the choice of a system *f*-number depends to a great extent upon the decision to use refractive or reflective, spherical or aspherical designs, as well as the monetary resources available to obtain high-quality optics. Once this decision is made, the layout design can proceed. However, a limiting-case design can be based upon diffraction limits as outlined in Chap. 13 and in the literature.[7]

12.5 ELECTRO-OPTICAL TELESCOPES

Many electro-optical systems utilize telescopes to maximize throughput and satisfy field-of-view (spatial) requirements. This section illustrates refractive and reflective designs.

12.5.1 Simple Refractive System

Perhaps the simplest telescope design is the system of Fig. 12.11, which consists of a single refractive lens and a detector mounted in the field stop. This system suffers from several problems: (1) chromatic effects, (2) chopper-induced noise, (3) infrared materials that are generally opaque in the visible and difficult to align and focus, and (4) a field of view subject to detector response nonuniformities.

The index of refraction is a function of both temperature and wavelength. This is especially inconvenient in cooled systems, where it is necessary to compensate for changes in temperature.

Most infrared systems utilize choppers (see Chap. 7) to convert the radiant flux to pulsating dc signals. The chopper must be very large if it is located in front of the lens of Fig. 12.11, and may introduce noise into the detector if it is located between the lens and the focal plane.

The alignment of cryogenically cooled infrared systems is very difficult when the lenses are fabricated from visually opaque materials. Generally, alignment can be verified only by a rather difficult calibration measurement; subsequent adjustments are made after the system has been brought to an ambient environment. Each cool-down, warm-up cycle often requires several days.

The system of Fig. 12.11 produces a sharp image on the detector field stop. This results in a field of view that exhibits the same nonuniformities as the detector itself.

12.5.2   Reflective System

The problems associated with the simple refractive system of Fig. 12.11 can be avoided by using reflective optics. Such a design layout, which makes use of a field lens, is illustrated as a refractive system first (Fig. 12.14). This arrangement solves the problem of chopper noise and nonuniform field of view by the introduction of a field lens L_2 located in the system field stop. Then, in Fig. 12.15 a similar reflective system is illustrated that solves the remaining chromatic and alignment problems.

The field stop of Fig. 12.14 is located in the focal plane of lens L_1, which corresponds to a telescope focused at infinity. The edge of L_1 is the limiting aperture and is therefore the entrance pupil of the system as in Fig. 12.13.

The field of view is determined by the chief ray, which is directed from the edge of the object to the center of the entrance pupil, and is given by Eq. (12.8). The field of view results from the convolution of the image of a point in the object plane, and the field stop, which for a small optical blur approaches the field-stop function (i.e., an ideal square function). The most efficient location for the optical chopper is in or near the focal plane because the optical field exhibits its smallest extent there, and in that location is sufficiently removed from the detector to overcome the noise problem.

The problem of the nonuniform field of view is also avoided by the use of the field lens that images the entrance pupil, rather than the object,

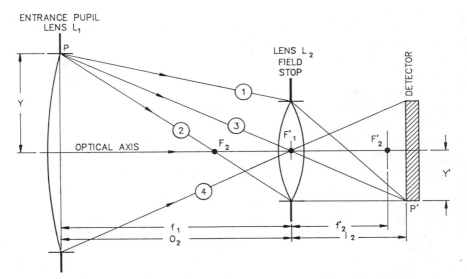

Figure 12.14   Illustration of a telescope designed so the entrance pupil is imaged upon the detector by lens 2.

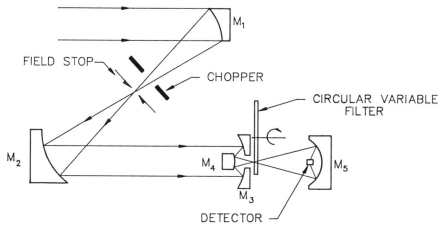

Figure 12.15 Illustration of an all-reflective system that utilizes a confocal off-axis system consisting of two parabolas, a Ritchey-Chrêtien relay, and a parabolic condenser.

upon the detector. This is illustrated in Fig. 12.14 as follows: Rays 1, 2, and 3 show that L_2 images the lens edge P on the detector edge P'. Rays 3 and 4 represent an on-axis ray bundle from a point on the object that irradiates the entire detector area. Thus, a ray from any point on the object will completely fill the detector, so the response everywhere within the field of view corresponds to the detector *average* response. The following is a numerical example of the design of the telescope of Fig. 12.14.

Example 4: Find the focal length of lens L_2 (Fig. 12.14) to produce an image of the entrance pupil on the detector.

Given: The focal length and height of L_1 are 20 and 7.5 units, respectively, and the field stop is located in the focal plane (the system object distance is infinity); the lens L_2 object distance is 20 units, and the detector and field-stop height is 1 unit (i.e., the detector and field stop are the same size).

Basic equations:

$$\frac{1}{i} + \frac{1}{f} = \frac{1}{o} \tag{12.4}$$

$$m = \frac{y'}{y} = \frac{i}{o} \tag{12.5}$$

Assumptions: Thin lens and paraxial rays

Solution: Lens L_2 images the entrance pupil on the detector, and by convention $y = +7.5$ and $y' = -1.0$. Thus, from Eq. (12.5)

$$m = -1.0/7.5 = -0.1333 = i/o$$

the object distance is -20 units and the image distance is

$$i = -0.1333 \times (-20) = 2.667$$

The focal length is, by Eq. (12.4),

$$f = \frac{1}{1/o - 1/i} = \frac{1}{-1/20 - 1/2.667} = -2.35$$

where the negative sign indicates that lens L_2 is a positive converging lens as illustrated in Fig. 12.1(a). Alignment is simplified in this design by making the detector somewhat oversized. ∎

The remaining problems associated with the simple system of Fig. 12.11, namely chromatic effects and difficulties in alignment associated with refractive systems, which make use of materials that are opaque in the visible, can be eliminated through the use of a reflective system.

Figure 12.15 illustrates an all-reflective system that utilizes a confocal off-axis system consisting of two parabolas, a Ritchey-Chrêtien relay, and a parabolic condenser. This system, designed for a circular-variable filter (sequential) spectrometer, produces three images, one for the field stop and chopper, one for the filter slit, and one for the detector.

The off-axis parabola M_1 serves as the collector and produces an image at the first field stop, which determines the field of view and is focused at infinity. The optical chopper is also located there. The second parabola M_2, provides collimated output. The confocal arrangement has the focal plane of M_1 and M_2 coincident and results in self-correction of aberrations.

The Ritchey-Chrêtien, M_3 and M_4, produces a second image that is used for the slit of a circular-variable interference filter. Finally, a third parabola M_5 collects the energy onto the detector, trading large solid angle for relatively small detector area.

12.5.3 Lyot System

A system of considerable interest (Sec. 14.4.2) is one used to reduce stray light caused by diffraction. Energy diffracted at the entrance aperture is

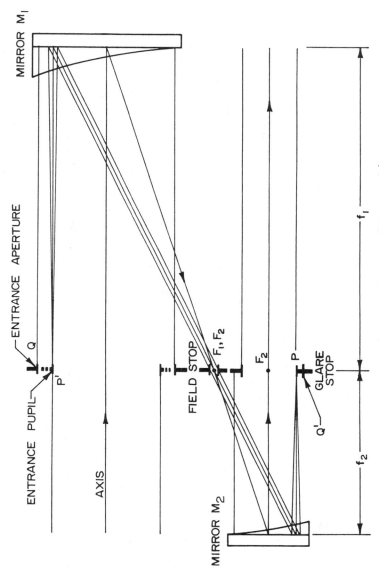

Figure 12.16 Illustration of a telescope designed to reduce stray light caused by diffraction at the entrance aperture through the use of a glare stop.

imaged somewhere in the system so that it can be blocked with a glare or Lyot stop. Such a system is illustrated in Fig. 12.16.

The field stop is located at the focal point of M_1, which results in the telescope being focused at infinity, and thus the edge of M_1 is an entrance aperture. The field stop is also located at the focal point of M_2, which results in parallel, or collimated, rays emanating from the system—these rays can be imaged upon a detector as shown in Fig. 12.14 or 12.15. The glare stop P is imaged at P'. The entrance aperture point Q is imaged at Q'; thus, any off-axis energy diffracted from the entrance aperture at Q is blocked by the glare stop at Q'.

The edge of mirror M_1, the entrance aperture Q, and the image of the glare stop (in object space, P') are all entrance pupils. According to the rule stated in Sec. 12.3, the system entrance pupil is the limiting aperture, or image of an aperture, in object space. The image of the glare stop is the limiting entrance aperture as viewed from a point on the object (at infinity) and functions as the system entrance pupil.

A detector condensing system, like that of Fig. 12.15, could be used after F_2 to concentrate the energy upon a detector. The entrance pupil can be imaged upon the detector (to avoid field-of-view problems), as in Fig. 12.14, by imaging the glare stop on the detector.

EXERCISES

1. A telescope objective lens has a focal length of 10 cm. Calculate the half-angle field of view and the solid-angle field of view for a 1-cm-diameter field stop.

2. A positive thin lens with a diameter of 2 cm and a focal length of 8 cm is located 20 cm from an object that extends from the optical axis to a height of 1 cm. Find the image distance and the magnification. Use graphic and numeric methods.

3. An ideal converging mirror has a focal length of 20 cm and is located 40 cm from an object that extends from the optical axis to a height of 3 cm. Find the image distance and the magnification. Use graphic and numeric methods.

4. A thin converging lens with an aperture 4 cm in diameter and a focal length of 8 cm has a 2-cm-diameter stop located 6 cm to the right of the vertex. Locate and identify the limiting aperture for object space as the entrance pupil for a point (*a*) at infinity and (*b*) 4 cm to the left of the vertex. Use graphic and numeric methods.

5. Prove that the solid-angle field of view, in units of steradians, is given by $\Omega = 4\sin^2\Theta$, where Θ is the half-angle of a square field of view.

6. Find the focal length of lens L_2 to produce an image of L_1 upon the 2-cm-diameter detector (Fig. 12.14), given that the focal length of $L_1 = 10$ cm, $y = 3$ cm, and the 1-cm-diameter field stop is located in the focal plane. Give a graphic and mathematical solution.

7. Design an off-axis system like that of Fig. 12.16 where numerical aperture is $F = 2$, the diameter of M_1 is 6 in., the full-angle field of view is 1×10^{-3} rad, M_2 diameter is 3 in., and the glare-stop diameter is 2.5 in. Find the field stop and entrance pupil diameters. Give a graphic and numeric solution.

REFERENCES

1. F. A. Jenkins and H. E. White, *Fundamentals of Optics*, McGraw-Hill, New York, 1957, p. 4.
2. Ibid., p. 6.
3. Ibid., p. 19.
4. Ibid., p. 46.
5. Ibid., p. 56.
6. Ibid., p. 57.
7. W. J. Smith, "Optical Systems," in W. L. Wolfe, Ed., *Handbook of Military Infrared Technology*, Office of Naval Research, Dept. of the Navy, Washington, DC, 1965, Ch. 10.

chapter 13

Imaging Systems

13.1 INTRODUCTION

This chapter is concerned with optical systems designed to create an image. The quality of an image formed by an optical system is determined by three factors: (1) the wave nature of light and resulting diffraction effects; (2) spherical and chromatic aberrations in the optical system, and (3) manufacturing defects.

Diffraction-limited systems represent a limit to the performance of any system of a given size. Consider a lens that forms an image of an incoherently illuminated narrow slit as illustrated in Fig. 13.1(*a*). The distribution of light in the image plane would be an exact duplicate for a hypothetical ideal system as shown in Fig. 13.1(*b*); however, this can never be realized because of the finite wavelength of light and the size of the system. The best that can be realized is the $(\sin x)/x$ distribution shown in Fig. 13.1(*c*). If there are aberrations, or manufacturing defects, present in the optical system, then the image may be very complex and could be of the form illustrated in Fig. 13.1(*d*).

The diffraction image of an incoherently illuminated slit is known as the *line-spread function*. If the slit is replaced by a point source, the corresponding diffraction-limited image is known as the *point-spread function*, and the image is referred to as the *airy disk* or optical blur.

The airy disk exhibits a cross section similar to that of Fig. 13.1(*c*) and contains 84% of the energy to the first dark rings for diffraction-limited

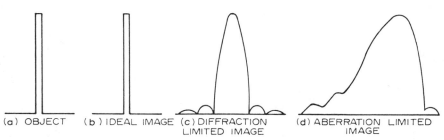

(a) OBJECT   (b) IDEAL IMAGE   (c) DIFFRACTION LIMITED IMAGE   (d) ABERRATION LIMITED IMAGE

Figure 13.1 Illustration of image of a narrow slit for ideal, diffraction-limited, and aberration-limited image.

conditions. The blur diameter θ (to the first dark ring) is given[1] by

$$\theta = 2.44\lambda/D \quad [\text{rad}] \tag{13.1}$$

where λ is the wavelength and D is the effective aperture (diameter of the entrance pupil). The Rayleigh criterion is that two adjacent equal-intensity point sources can be considered resolved if the first dark ring pattern falls on the center of the other pattern. Thus, the angular resolution for the Rayleigh criterion is half that given by Eq. (13.1).

The design problem consists of quantitatively specifying an optical system consisting of lenses, or other imaging devices, and detectors and signal-conditioning circuits to obtain the *system performance objectives*. However, the resolving power of an optical system as given by the Rayleigh and similar criteria is only a partial description of system performance.

The modulation transfer function (MTF) theory of a sensor provides a tool that can be used by an optical designer to quantitatively analyze system performance and to determine how each element in the system is contributing to that performance.[2, 3]

The MTF method is analogous to that used with electric networks in which the network is stimulated with an electric signal. The relationship between the input and output signals can be used to describe the transfer function of the network. In this analogy, the object scene, like a complex electric signal, contains patterns that are referred to as "spatial frequencies." The system frequency response is obtained, as with electrical networks, by considering its response to a particular object scene distribution. For example, the response to an incoherent point source, the point-spread function, is analogous to an impulse response function in electrical network theory. The response to a "knife-edge," which is an abrupt transition from a dark area to a light area, is analogous to a step function response in electrical network theory.

It is also possible (although difficult) to construct a one-dimensional sine-wave source. In this case a series of alternate bright and dark bars with a sine-wave distribution is constructed as shown in Fig. 13.2. A cycle includes the dark and light regions, and the spatial frequency can be

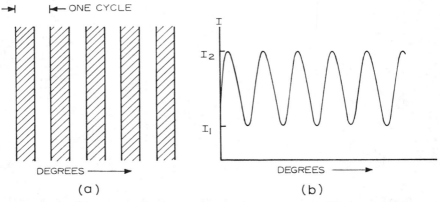

Figure 13.2 Object scene modeled with (a) a bar pattern and (b) the corresponding sine-wave distribution.

specified in terms of cycles per degree (or radian) or, for a specific system, in cycles per millimeter in the focal plane.

The quantitative response of the optical system's electric output signal to such a spatial scene as that given in Fig. 13.2 is evaluated in terms of the modulation transfer function as follows.

It is known from Fourier transform theory[4] that an object scene consisting of a one-dimensional sine-wave distribution [as shown in Fig. 13.2(b)] results in a sine-wave distribution in the image plane of a diffraction-limited aperture. Provided the detector is linear, and neglecting the effects of noise, the electric output signal is also sinusoidal. The only effect of the system is to reduce the contrast or modulation.

For equal crests and troughs, the modulation is defined as the relative deviation from the average, which for the case of Fig. 13.2(b) is given by

$$\text{Modulation (object scene)} = \frac{I_2 - I_1}{I_2 + I_1} \qquad (13.2)$$

The modulation of the sensor output voltage in response to the object scene is

$$\text{Modulation (output voltage)} = \frac{V_2 - V_1}{V_2 + V_1} \qquad (13.3)$$

The modulation transfer function is an expression of the ability of the system to faithfully reproduce the object scene modulation and is given[5] by

$$\text{MTF} = \frac{\text{modulation (output voltage)}}{\text{modulation (object scene)}} \qquad (13.4)$$

The MTF is a function of spatial frequency and for a perfect system has a value of unity. It is also a function of wavelength and field of view. It may

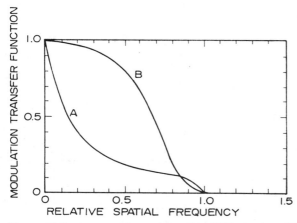

Figure 13.3 A comparison of possible MTFs for two different lenses that exhibit the same resolution.

have a different value perpendicular to the direction of scan than in the scan direction, or it may vary off-axis from what it is on-axis.

The utility of the MTF function is illustrated in Fig. 13.3, which gives possible modulation transfer functions for two different optical systems that exhibit the same resolution based upon the Rayleigh criterion. Lens A is superior for high spatial frequencies, while lens B is best for low spatial frequencies.

Designing transfer functions like those illustrated in Fig. 13.3 is analogous to electrical filter theory in which equalization networks are designed to obtain a particular frequency emphasis.

13.2 SYSTEM MTF

The system MTF results from the system components[6] as shown in Fig. 13.4. This development concentrates on the effects of diffraction, aberrations, and the detector scanning aperture before noise and nonlinearities are introduced.

One of the convenient features of the MTF method is that the system MTF can be obtained simply as the product of the optical MTF and the detector scanning aperture MTF.

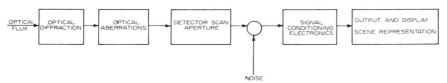

Figure 13.4 Major contributors to system modulation transfer function in a typical imaging system.

13.3 OPTICAL MTF

The MTF of the optical system is the ratio of the modulation of the image scene to that of the object scene. In principle, it could be measured by scanning the image plane with an extremely fine slit and associated detector. Such a measurement is generally difficult to make. It requires fabrication of a sine-wave scene of known contrast and a very sensitive detector that can detect the energy in the image plane when limited by a slit narrow enough to have negligible effect on the optical system MTF.

The MTF of a corrected, or ideal, optical system is diffraction-limited and is dependent upon the wavelength of light used and the effective diameter of the collecting aperture. Systems designed to produce a scene representation are generally optimized so that diffraction theory can be applied as follows: The MTF for a clear circular aperture is given[2] by

$$\text{MTF} = \frac{2}{\pi}(\phi - \cos\phi\sin\phi) \tag{13.5}$$

where

$$\phi = \arccos(\lambda FN) \tag{13.6}$$

where λ is the wavelength, F is the relative aperture (*f*-number), and N is the spatial frequency.

It is evident that MTF is zero when ϕ is zero. Thus, the "limiting frequency" N_L is given in cycles per millimeter by

$$N_L = \frac{1}{\lambda F} \quad [\text{c/mm}] \tag{13.7}$$

and is the frequency for which the MTF = 0. Thus, Eqs. (13.5) and (13.6) can be solved in terms of the relative frequency N/N_L since

$$\phi = \arccos(N/N_L) \tag{13.8}$$

Figure 13.5 gives the solution to Eq. (13.5) in terms of relative frequency.

The limiting frequency, Eq. (13.7), can be expressed in cycles per radian by multiplying by the effective focal length, and is given by

$$N_L = D/\lambda \quad [\text{c/rad}] \tag{13.9}$$

Example 1: Find the MTF for a diffraction-limited optical telescope for a spatial frequency of 1.0 c/mrad at 10-μm wavelength.

Given: The effective collector diameter is 10 cm.

Basic equations:

$$\text{MTF} = \frac{2}{\pi}(\phi - \cos\phi\sin\phi) \tag{13.5}$$

$$\phi = \arccos(N/N_L) \tag{13.8}$$

$$N_L = D/\lambda \quad [\text{c/rad}] \tag{13.9}$$

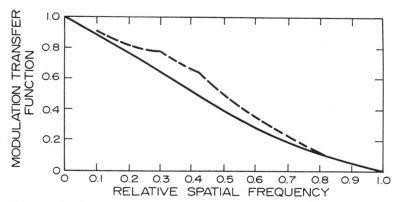

Figure 13.5 Modulation transfer function for a circular aperture resulting from optical diffraction. The dashed line is the MTF for a square-wave target.

Assumptions: Ideal diffraction-limited, i.e., no spherical aberrations or manufacturing defects.

Solution: The limiting spatial frequency of the telescope is given by

$$N_L = \frac{D}{\lambda} = \frac{10[\text{cm}] \times 10^{-2}[\text{m/cm}]}{10 \times 10^{-6}[\text{m}]} = 1 \times 10^4 \text{c/rad}$$

The relative spatial frequency for 1.0 c/mrad is

$$\frac{N}{N_L} = \frac{1[\text{c/mrad}]}{1 \times 10^4[\text{c/rad}] \times 10^{-3}[\text{rad/mrad}]} = 0.1$$

Thus,

$$\phi = \arccos(N/N_L) = \arccos(0.1) = 84.26° = 1.47 \text{ rad}$$

and

$$\text{MTF} = \frac{2}{\pi}(1.47 - \cos 84.26° \sin 84.26°) = 0.87$$

which agrees with Fig. 13.5 for a relative frequency of 0.1.

Note: In Example 1 the limiting frequency could also be given in cycles (or lines) per millimeter as is common in the lens industry. This is obtained using Eq. (13.7) provided the focal length f is known; then $F = f/D$. ■

The usual "resolution limit" defined by the Rayleigh criterion, as given by half that of Eq. (13.1), might be interpreted to imply that the system is useful to scene frequencies given by $0.82 N_L$; that is,

$$\theta = 1.22\frac{\lambda}{D} = \frac{1.22}{N_L} = \frac{1}{0.82 N_L}$$

Using the method of Example 1, the MTF is found to be 0.089, which means that the system reproduces about 9% of the original scene contrast. Such low values of scene contrast may be discernible, but for some applications they can hardly be considered faithful reproduction.

13.4 SCANNING DETECTOR MTF

A linear rectangular detector array exhibits a modulation transfer function with a $(\sin x)/x$ distribution[7] as shown in Fig. 13.6. The meaning of negative values of MTF is that the dark and light regions are reversed in position in the output. The abscissa in Fig. 13.6 is given in terms of relative frequency as before. The array provides a sample set of the image scene obtained at the frequency N_s with units of detectors/mm or detectors/rad. One cycle of the scene is equal to the detector width when the scene frequency is equal to the sample frequency. In this case, the detector responds to the average value only, and consequently the MTF $= 0$; this is analogous to the limiting frequency for a lens, N_L. The Nyquist frequency is defined as $N_s/2$, the frequency beyond which no useful information is obtained.[8] This corresponds to the sampling theorem criteria and relates to aliasing errors as given below.

The instantaneous field of view of a detector element when coupled to a lens system is obtained as the ratio of the detector pitch w to the effective focal length and is expressed in radians. Thus the sampling frequency in cycles per radian is

$$N_s = f/w \quad [\text{c/rad}] \tag{13.10}$$

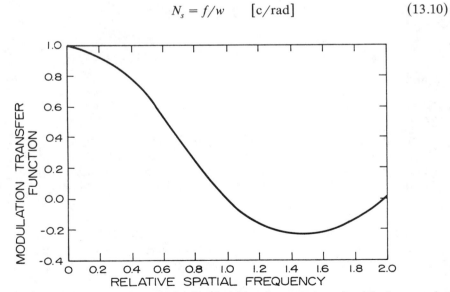

Figure 13.6 Modulation transfer function for a linear rectangular detector scanning aperture or linear array.

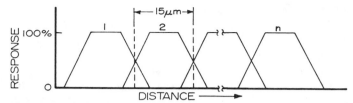

Figure 13.7 Detector array response profile.

where f is the effective focal length and w is the detector pitch defined as the detector center-to-center distance (see Fig. 13.7).

Example 2: Find the MTF for a linear rectangular scanning detector array, when used with a lens, for an image frequency of 1000 c/rad.

Given: The detector pitch is 15 μm, and the effective focal length of the lens is 55 mm.

Basic equation:

$$N_s = f/w \quad [\text{c/rad}] \tag{13.10}$$

Assumptions: None

Solution: The detector sampling frequency expressed in cycles per radian is given by

$$N_s = \frac{f}{w} = \frac{55[\text{mm}]}{15[\mu\text{m}] \times 10^{-3}[\text{mm}/\mu\text{m}]} = 3666.7 \text{ c/rad}$$

The relative frequency for the detector is given by

$$N/N_s = 1000[\text{c/rad}]/3666.7[\text{c/rad}] = 0.27$$

An analytical solution for the MTF is not convenient in this case. However, by Fig. 13.6, the MTF is approximately 0.8. ∎

13.5 SYSTEM MTF

The system MTF is given as the product of the optical system and the detector MTF. The system MTF can be specified in a way that relates to the system performance criteria. The following example illustrates. An astronomical system is being designed to resolve two adjacent stars. The stars are considered resolved if the signal response exhibits a minimum between the two point images. However, sensitivity estimates indicate that a signal-to-noise ratio of about 10 is likely. In this case the dip is approximately equal to the rms noise for an MTF of 0.1 assuming the object contrast is 100%; then the possibility of detecting the dip is rather poor. On the other hand, an MTF of 0.5 yields a dip of at least 5 standard deviations and can be detectable with high probability. For a given probability there exists a trade-off between MTF and SNR, and a quantitative evaluation of the effect of MTF is made in relationship to the system noise and the probability of detecting signals of the appropriate spatial frequencies.

It has been suggested for remote sensing[9] that the noise equivalent differential reflectance (NEdR), which is defined as the change in scene reflectance that produces a change in detected signal equivalent to the instrument noise, be no greater than 0.5%. This corresponds to SNR = 200. The information is contained in the modulation rather than the peak scene signal V_m; thus, we can write

$$\frac{V_n}{V_m \, \text{MTF}} = \frac{1}{\text{SNR} \times \text{MTF}} = \text{NEdR} \qquad (13.11)$$

where SNR is defined for a scene exhibiting low spatial frequencies (i.e., MTF = 1).

13.6 MEASUREMENT OF MTF

Diffraction-limited MTF provides a standard of excellence of design and fabrication as well as an evaluation of the limiting performance of any given size system. However, most systems will fall short of this ideal because of aberrations, scattering, and other imperfections. It is therefore usually necessary to measure a system MTF.

There exist a relatively large number of methods by which the MTF can theoretically be measured.[10] However, all such measurements, though simple in principle, are difficult to make.

The resolution of commercial lenses is often measured by photographing bar patterns of various spatial frequencies. The resolution corresponds to the highest frequency for which the resulting film bar pattern can be perceived with a microscope. It depends not only on the quality of the lens but also on the film used in the test. It is likely that such a test corresponds to a system MTF of about 0.1, which is close to the ability of the human eye to detect changes in intensity.

Examination of the image on a film produced by photographing a scene pattern yields a measure of the *system* MTF that is the product of the optical MTF and the film MTF. It might be argued that a measurement of the system MTF is best, since that is the way the system is to be used. However, knowledge of the optical subsystem and the detector subsystem MTFs is required before the optical designer can optimize each contributor to system performance.

The method described above, of stimulating the system by scanning an object scene consisting of alternate bright and dark bars of sine-wave distribution of known contrast, is the most straightforward in principle. However, spatial sine-wave distributions of known contrast are not readily available. A more convenient pattern is the square-wave target, which is a pattern of alternate bright and dark bars of equal width.[11] In general, the MTF is higher for a square wave than for a sine-wave distribution (see Fig. 13.5). For example, the MTF for a diffraction-limited system at relative frequencies between 0.25 and 0.5 is about $+0.1$ greater for a square wave than for a sine wave.

Another method makes use of a knife-edge test and a scanning detector array.[12] This is analogous to testing an electrical network with a step function rather than a sine wave and should yield the same results provided the system does not exhibit overshoot, i.e., provided it is critically damped or overdamped.

A simple way to perform the knife-edge test for a system consisting of a lens and linear detector array is to image the system on a razor blade that is back-illuminated with a collimated light source. The video output of the signal-processing system is displayed on an oscilloscope. The dark-to-light transition appears as a step-function transition. The rise time T_r is observed in terms of the number of detector elements required to complete the 10 to 90% rise time. Thus,

$$T_r = \frac{N}{N_s} \quad \text{[rad]} \tag{13.12}$$

where N is the number of detector elements for the 10 to 90% rise time and N_s is detector sampling rate in detectors per radian. The value of N_s is given by Eq. (13.10), where one detector width equals 1 cycle.

The MTF is obtained for the Nyquist frequency (i.e., for a scene frequency of $N_s/2$) by

$$\text{MTF}|_{N_s/2} = 1/N \tag{13.13}$$

Thus, the MTF can be obtained for the frequency $N_s/2$.

Example 3:   Measure the MTF of a Pentax 55mm Takamar $f/2$ lens using a collimated beam-expanded laser and a detector array of 15 μm pitch.

Given: The knife-edge test is used with a beam-expanded 6328-A laser. The 10 to 90% rise time is observed as 1.8 elements.

Assumptions: None

Solution: The Nyquist frequency, for which the knife-edge test yields the MTF, is one-half the sampling frequency:

$$\frac{N_s}{2} = \frac{f}{2w} = \frac{1}{2} \left(\frac{55[\text{mm}]}{15[\mu\text{m}] \times 10^{-3}[\text{mm}/\mu\text{m}]} \right)$$
$$= 1833.3 \text{ c/rad}$$

By Eq. (13.13) and the observational data,

$$\text{MTF} = 1/N = 1/1.8 = 0.56.$$

for a spatial frequency of 1833 c/rad.

The spatial frequency of 1833 c/rad can be described in terms of cycles (lines) per mm as follows:

$$\frac{N_s}{2} = \frac{N\,[\text{c/rad}]}{f\,[\text{mm/rad}]} = \frac{1833}{55} = 33 \text{ lines/mm}$$

where f is the focal length of the lens expressed in mm/rad. ∎

The method of Example 3 can be used with a single detector rather than an array. The modulation transfer function is the same for a single *scanning* detector as it is for the array, provided the single detector is the same size as the array detectors. The measurement technique is as follows: The scanning detector is physically scanned so that it sequentially occupies adjacent positions, as in the equivalent array, and the output is recorded for each position. This can be easily accomplished on a storage scope, and the pattern so produced is, in principle, identical to that obtained with the array of Example 3. An alternative method is to move the object so the image scans the detector.

13.7 ALIASING

The linear scanning detector or linear detector array produces a sample set. The sampling theorem (see Chap. 7) states that a continuous signal must be sampled at a rate twice that of the highest frequency component in order to faithfully reconstruct the original signal. Aliasing is a form of error that occurs when the signal contains frequencies that exceed this maximum frequency,[13] which is referred to as the Nyquist frequency.

The detector sample frequency must be at least twice the maximum scene frequency. Usually, the sample frequency is set at 5 times the maximum scene frequency, since the MTF is not an ideal low-pass filter. Since the limiting frequency N_s for the detector is the frequency for which the MTF is zero, the sampling criterion guarantees a relative frequency of 0.2 (for a factor of 5), which yields a detector MTF of approximately 0.9.

The spatial frequencies contained in an object scene depend upon the amount of detail in that scene. Natural scenes tend to have a great deal of energy in the high-frequency part of the spectrum. The optical system, because of diffraction and aberrations, behaves as a spatial filter performing certain smoothing functions. The MTF is a representation of the normalized frequency response of that spatial filter.

Aliasing errors occur when the lens fails to filter out frequencies higher than the maximum rate. A designer may specify an MTF, utilizing spatial filtering techniques, that is optimized for a particular application. In any case, a compromise is called for, since reducing the MTF at frequencies above the maximum target frequency may also reduce the response at the target frequency.

13.8   DESIGN OF MTF—A SPECIFIC EXAMPLE

A remote-sensing system is designed to detect and count deer to provide a tool for wildlife management. The objective is to utilize a detector array with appropriate optical collector and signal-conditioning electronics sub-systems to survey the deer from an airborne platform at a height of 1500 ft.

The information useful for deer classification is contained in the modulation of the scene flux. A nonunity value of MTF implies a loss of information and is manifest as a decrease in the SNR or the differential reflectance and a greater probability of misclassification. The criteria given above of NEdR = 0.5% and Eq. (13.11) are appropriate.

The near instantaneous field of view (IFOV) is given by IFOV = 1/1500 rad for a 1-ft-wide deer at 1500 ft. The deer frequency is

$$N_d = \frac{1}{2 \times \text{IFOV}} = 750 \text{ c/rad}$$

The Reticon[14] linear detector array has 1728 elements with an aperture (height) of 16 μm and a width of 15 μm. The detector response function overlaps with a trapezoidal pattern (see Fig. 13.7). The detector sampling frequency (for which MTF = 0) for the array is given by Eq. (13.10). The maximum scene frequency permitted by the sampling theorem is half that given by Eq. (13.10); however, a factor of $\frac{1}{5}$ is used (as discussed above), rather than $\frac{1}{2}$, to avoid aliasing errors. The focal length f must be chosen to set the sampling frequency of the array at 5 times the deer frequency, and thus, by Eq. (13.10),

$$f = N_s w = 5 \times 750 \times 15[\mu\text{m}] \times 10^{-3}[\text{mm}/\mu\text{m}] = 56.25 \text{ mm}$$

Each element has an IFOV of

$$\Delta\theta = 15[\mu\text{m}]/56[\text{mm}] = 0.25 \text{ mrad}$$

(assuming the lens is perfect), which at 1500 ft yields a footprint of

$$0.25 \times 10^{-3}[\text{rad}] \times 1500[\text{ft}] = 0.41 \text{ ft}$$

The relative frequency of the array at the deer frequency of 750 c/rad is

$$\frac{N}{N_s} = \frac{750 \times 15[\mu\text{m}] \times 10^{-3}[\text{mm}/\mu\text{m}]}{56[\text{mm}]} = 0.20$$

The detector MTF is obtained from Fig. 13.6 as about 0.93.

The following is a calculation to determine the diffraction-limited aperture to satisfy the above requirements: The relative frequency for an ideal diffraction-limited circular aperture of MTF = 0.93 is approximately $N/N_L = 0.05$ from Fig. 13.5. The limiting frequency N_L is obtained for the deer frequency as

$$N_L = \frac{750[\text{c/rad}]}{0.05} = 1.5 \times 10^4 \text{ c/rad}$$

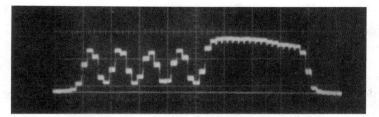

Figure 13.8 Oscillogram of the electrical response of a Reticon HB 1728 detector array and Pentax 55 mm Takamar $f/2$ lens to a square-wave bar pattern of 750 c / rad.

The aperture diameter for an ideal diffraction-limited system is given by Eq. (13.9) as

$$D = \lambda N_L = 0.66[\mu m] \times 10^{-3}[mm/\mu m] \times 1.5 \times 10^4 [rad/c] = 9.9 \text{ mm}$$

The lens f-number is

$$F = f/D = 55/9.9 = 5.5$$

where f is the effective focal length and D is the effective diameter of the aperture.

The camera lens of Example 3 is considered as a possible candidate for this system since it has a focal length and f-number close to that required. The system is equipped with a narrow-band (0.02 μm) interference filter, so the flux is relatively monochromatic.

A high-resolution collimator and associated reticle are used to produce a square-wave bar pattern corresponding to the deer frequency, focused at infinity. The output signal is observed on an oscilloscope as shown in Fig. 13.8, which shows the response to four cycles of frequency 750 c/rad and to a large (low-frequency) bright area.

The large bright area in Fig. 13.8 gives the object scene contrast (modulation), since it corresponds to low spatial frequencies for which the system MTF approaches unity; the modulation for the 750-c/rad square wave is observed to be somewhat less. Careful examination of the oscillogram yields a measured system MTF of about 0.70; thus, the lens MTF = 0.7/0.93 = 0.75, which corresponds to 13.6 c/mm in the lens focal plane.

Using the criteria NEdR = 0.5% and Eq. (13.11), we have

$$\text{SNR} = \frac{1}{\text{NEdR} \times \text{MTF}} = \frac{1}{0.005 \times 0.7} = 286$$

as the required SNR for a uniform scene.

EXERCISES

1. Find the MTF at a frequency of 20 c/mrad for a 20-in.-diameter (aperture) optical telescope that is diffraction-limited at a wavelength of 10 μm.
2. The SNR for very low spatial frequencies is 25. Find the SNR for a 20-c/mrad spatial frequency of a diffraction-limited 10-in.-diameter (aperture) telescope at 10 μm.

3. A scanning radiometer is to be designed that is capable of detecting targets 0.2 m in diameter at a range of 1000 m. Radiometric calibrations predict a signal-to-noise ratio of 100 for a sensor with MTF = 1. Design a diffraction-limited telescope that provides an SNR of 10 at a wavelength of $\lambda = 10$ μm at the target spatial frequency. *Hint:* Model the problem in terms of a periodic function where target is $\frac{1}{2}$ cycle.

4. The SNR for a system exhibiting an MTF of 1 is 100. Find the SNR for a spatial frequency of 20 c/mrad for a diffraction-limited 12-in.-diameter telescope at 10 μm.

5. Find the MTF at a frequency of 930 c/rad for a 2-cm-diameter camera lens at 1.0 μm. Assume diffraction-limited conditions.

6. Given that a photographic bar chart produces 4000 c/rad, find the number of lines per millimeter (cycles per mm) in the focal plane of a 55-mm focal length camera lens.

REFERENCES

1. W. L. Wolfe and G. J. Zissis, Eds., *The Infrared Handbook*, Office of Naval Research, Dept. of the Navy, Washington, DC, 1978, p. 8-28.
2. W. J. Smith, *Modern Optical Engineering*, McGraw-Hill, N.Y., 1966, p. 311.
3. F. Abbott, in *Optical Industry & Systems Encyclopedia & Dictionary*, Optical Publishing, Pittsfield, MA, 1979, p. E-289; also published in *Optical Spectra* (March, April, June, July, 1970).
4. Ibid., p. E-295.
5. W. L. Wolfe and G. J. Zissis, Eds., *The Infrared Handbook*, Office of Naval Research, Dept. of the Navy, Washington, DC, 1978, p. 8-31.
6. H. B. Barhydt, D. P. Brown, and W. B. Dorr, "Comparison of Spectral Regions for Thermal Imaging Infrared Detectors," *Proc. Infrared Information Symp.*, **14**, 12 (August 1970).
8. W. L. Wolfe and G. J. Zissis, Eds., *The Infrared Handbook*, Office of Naval Research, Dept. of the Navy, Washington, DC, 1978, p. 12-18.
9. J. B. Wellman et al., "Imaging Spectrometer Technologies for Advanced Earth Remote Sensing," *Proc. Soc. Photo-Opt. Instrum. Eng.*, **345**, 32 (May 1982).
10. F. Abbot, in *Optical Industry & Systems Encyclopedia & Dictionary*, Optical Publishing, Pittsfield, MA, 1979, pp. E297–E303.
11. W. L. Wolfe and G. J. Zissis, Eds., *The Infrared Handbook*, Office of Naval Research, Dept. of the Navy, Washington, DC, 1978, p. 8-5.
12. R. Hopwood, "Design Considerations for a Solid-State Image Sensing System," paper submitted to the Society of Photo-Optical Instrumentation Engineers for SPIE's Technical Symposium East, April 1980, p. 8.
13. Y. Talmi and R. W. Simpson, "Self-Scanned Photodiode Array: A Multichannel Spectrometric Detector," *Applied Optics*, **19**, 1401–1414 (May 1980).
14. EG & G Reticon FL 1728 H silicon monolithic self-scanning linear photodiode array. (EG & G, 345 Potrero Ave., Sunnyvale, CA 94086.)

chapter *14*

Baffling in Optical Systems

14.1 INTRODUCTION

Baffling in optical systems is used to reduce the propagation of radiant flux by scattering from nonoptical components.[1] The unwanted flux, referred to as "stray light,"[2] often results in *spatial* and/or *spectral impurities* in sensor measurements.[3]

The reduction of stray light in optical systems relates to the *quality* of the measurements achieved using a sensor. The sensor aperture is bombarded with unwanted flux that arrives from spatial regions outside the sensor field of view, such as the sun, earth, and lights. The sensor output is also bombarded with unwanted flux that is out-of-band, that is, outside the spectral band of interest. The sensor output, for *spatially* and *spectrally pure* measurements, is independent of these unwanted forms of flux. Therefore, the reduction of stray light in optical systems improves the quality of the resultant measurements.

There are two basic types of baffles used in optical telescopes; they can be classified as *imaging* and *nonimaging*.

A "sunshade" baffle is typical of nonimaging baffle systems for which the unwanted source is far enough off-axis that direct illumination of the optical surfaces can be avoided by placing baffles ahead of the optics.

Glare stops, placed at the image of the system entrance pupil, are used in systems that require rejection of unwanted sources located very near the optical axis when direct illumination of the optical surface cannot be avoided.

Low-scatter optical surfaces are used when direct illumination of these surfaces by unwanted off-axis sources cannot be avoided. An example is the solar coronagraph in which the direct radiation of the sun is imaged and blocked by a glare or Lyot stop, so that the brighter parts of the corona can be measured at the experimenter's convenience rather than waiting for a solar eclipse.

Sunshade baffles, glare stops, and low-scatter surfaces are combined in systems to achieve the ultimate in stray-light rejection; each of these devices is considered in this chapter.

14.2   THE nTH-ORDER BAFFLE

The nth-order baffle is a useful concept to illustrate the design of baffle systems in general. The nth-order baffle system is defined as one for which the stray radiation has been successively scattered from a series of n baffles so that after the nth successive scattering the radiation has been reduced by some coefficient raised to the nth power. Such a system is appropriately called an nth-order baffling system.

A third-order baffling system is illustrated in Fig. 14.1. Radiation entering the system at an angle greater than or equal to α relative to the optical axis can fall upon the first-order baffle only. Scattered radiation from this baffle can fall upon the second-order baffle only, and scattered radiation from the second can fall only upon the third-order baffle; hence the stray radiation incident upon the detector can come only from the third-order baffle.

The analysis of the nth-order baffle is accomplished by examining the effects of scattering of radiant energy entering the aperture of a sensor. It is

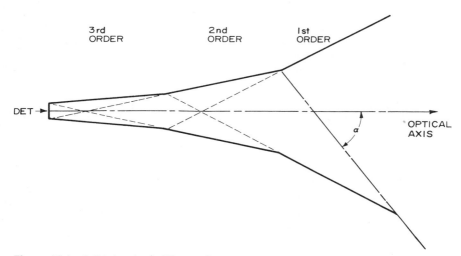

Figure 14.1   A third-order baffling system.

assumed that the baffle surfaces are treated so that they absorb a large fraction of the incident radiation.

If the stray radiation originates at a source radiating L_0 W cm^{-2} sr^{-1}, the total power entering the baffle is

$$\Phi_T = L_0 A_b \omega_b \cos \theta \quad [\text{W}] \quad (14.1)$$

where A_b is the baffle aperture area and ω_b is the solid angle subtended by the source at the baffle (in the direction θ). The energy is incident only upon the first-order baffle (see Fig. 14.1) provided the source is located at $\theta \geq \alpha$.

Of the power incident upon the first-order baffle, a fraction Γ_1 is scattered to the second-order baffle (where Γ is an empirical parameter determined experimentally for a given baffle configuration).

If this process is repeated on n successive surfaces, the power incident upon the detector from the nth-order baffle is

$$\Phi_d = \Gamma_1 \Gamma_2 \cdots \Gamma_n \Phi_T = \Gamma^n \Phi_T \quad (14.2)$$

where it is assumed that the scattering coefficient is similar for each baffle.

The power incident upon the detector when the source is on-axis ($\cos \theta = 1$) is given by

$$\Phi_0 = L_0 A_c \omega_c \quad [\text{W}] \quad (14.3)$$

where A_c is the collector area and ω_c is the solid-angle field of view.

The relative response, $\mathscr{R}_b(\theta)$, is defined as the off-axis response to a point source. The relative response for the nth-order baffle is the ratio of the off-axis power to the on-axis power:

$$\mathscr{R}_b(\theta) = \Gamma^n \frac{\Phi_d}{\Phi_0} = \frac{\Gamma^n A_b \omega_b \cos \theta}{A_c \omega_c} \quad (14.4)$$

The performance of such a baffle is illustrated in Fig. 14.2, which is the measured response to a point source of a system with a sunshade baffle. A layout drawing of the baffle is given in Fig. 14.3.

The data plotted in Fig. 14.2 are interpreted as follows: The rapid drop in the response at 2° off-axis results from the point-source image convolving with the edge of the system's first field stop.

The inclined region from 2 to 12° results from illumination of the primary mirror, by the point source, from which radiant energy scatters into the field stop.

The rapid drop at 12° results from the shielding of the primary mirror by the baffle.

The inclined region from 12 to 55° results from the illumination of the knife-edges. The radiant energy scatters from the knife-edges to the primary mirror, from which it scatters into the field stop.

The rapid drop (into system noise) at 55° results from the shielding of the knife-edges by the forebaffle.

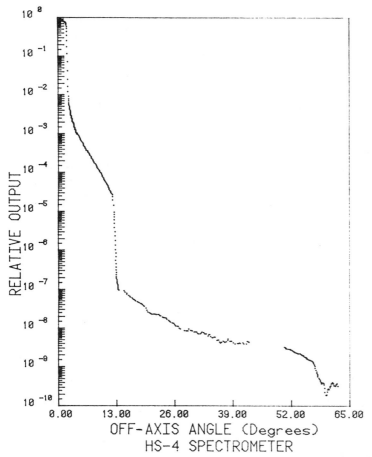

Figure 14.2 Measured off-axis response to a point source of the USU spectrometer baffle.

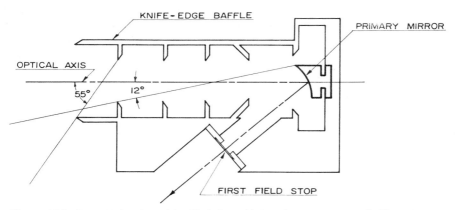

Figure 14.3 Layout drawing of the Utah State University spectrometer baffle.

These scattering coefficients are typical of those obtained for baffles and standard optical quality mirrors. The principle of the nth-order baffle illustrated in Fig. 14.1 applies in the system of Fig. 14.3 as follows: The sunshade is the first-order baffle. Energy entering the aperture at angles greater than α (12° in this case) is incident upon the sunshade only. The scattering coefficient Γ_1 is observed in Fig. 14.2 to be about 1×10^{-4}. The mirror functions as the second-order scatterer in this system; its scattering coefficient is about 1×10^{-3}, as observed in Fig. 14.2.

The overall performance depends upon the location of the off-axis source. Between 2 and 12° the system functions as a first-order baffle with $\mathscr{R}_1 \cong 1 \times 10^{-3}$ (the function $\cos \theta$ in Eq. (14.1) is very nearly unity for all cases). The system functions as a second-order baffle for angles of 12 to 55° and $\mathscr{R}_2 \cong 1 \times 10^{-7}$. Finally, for angles greater than 55°, energy does not enter the baffle directly from the source.

14.3 KNIFE-EDGE BAFFLES

The coefficient of scattering for the sunshade baffle is improved by incorporating knife-edges in the design. The baffle surface is constructed with many radiation-trapping cavities formed by thin-walled projections as illustrated in Fig. 14.4. Any radiation incident upon the cavity is completely absorbed after many successive reflections. The deeper the cavities, the more complete is the resultant absorption of incident radiation.

The scattering coefficient is proportional to

$$\Gamma \cong A_2/A_1 \tag{14.5}$$

where A is the area per unit length. The ratio A_2/A_1 is the fraction of the total incident radiation that is available for scattering on the edges of the partitions. In practice, these partitions are manufactured with sharp or knifelike edges to reduce the scattering area. In some cases these partitions are honed to a razorblade-like edge.

A design goal for knife-edge baffles is to make the cavities as deep as overall space limitations will permit in order to minimize the number of scattering edges. The following is a rationale for designing a sunshade baffle as illustrated in Fig. 14.5. Note: The item numbers (1 to 9) correspond to the construction lines in Fig. 14.5.

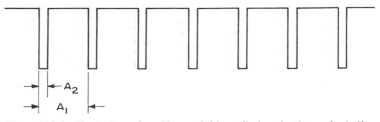

Figure 14.4 Illustration of cavities and thin-walled projections of a knife-edge baffle.

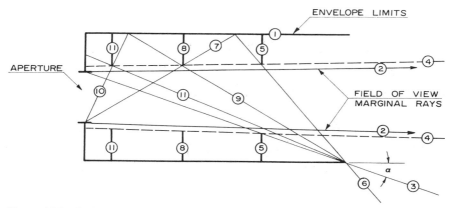

Figure 14.5 Rationale for the design of a knife-edge baffle based upon a fixed space limitation (1) and a given aperture.

1. Draw the physical limits for the system.
2. Draw the marginal rays from the edge of the entrance aperture.
3. Draw the baffle angle (primary angle) from the upper entrance aperture to the lower right physical limit. (The off-axis source must be located at or beyond α to prevent direct illumination of the entrance aperture.)
4. Draw the knife-edge limits to ensure that baffle edges are out of the field of view.
5. Locate the first knife-edge so its edge is at the intersection of the knife-edge limit (4) and the primary baffle construction line (3). (This knife-edge partition shields the aperture from any stray light scattered from the walls to the right of the partition.)
6. Draw the scattering ray from the first knife-edge to the lower right physical limit.
7. Draw the lower-limit ray to intersect with the scattering ray (6) on the wall.
8. Locate the second knife-edge at the intersection of the knife-edge limit (4) and the lower-limit ray (7). [This construction provides the greatest separation of knife-edges and yet shields the aperture from any direct scattering off the walls (1).]
9. Repeat the above procedure to locate the third baffle (11), using construction lines 9 and 10. (Line 11 shows that knife-edge 11 is closer to the aperture than necessary. However, this procedure proves that for the physical limits given, a maximum of three knife-edges are required. It is now possible to reposition the knife-edges to achieve more uniform location consistent with the objective of preventing any scattering from the wall from entering directly into the aperture.)

A major objective of the sunshade baffle is to prevent direct illumination of the optical components located in the entrance aperture (Fig. 14.5).

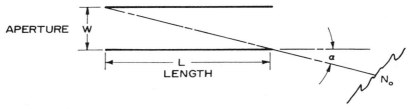

Figure 14.6   Limitations to the sunshade baffle based upon L/W ratio.

The limits to the achievement of this objective in the design of a sunshade baffle are the field of view and the primary baffle angle. This limitation can be described by consideration of a single-baffle system as illustrated in Fig. 14.6.

Neglecting the dependency on field of view (which is generally small), the limitation to a baffle design can be expressed in terms of the ratio L/W, where

$$L/W = \cot \alpha \qquad (14.6)$$

This leads directly to the conclusion that baffling is difficult for large apertures and small α.

The next section considers diffraction and low-scatter optics, which are used when direct illumination of the primary collector (entrance aperture, Fig. 14.4) cannot be avoided.

14.4   IMAGING SYSTEMS

There are two problems associated with the rejection of stray light when the unwanted source is within a degree or so of the optical axis. First, the primary collecting optic is directly illuminated by the unwanted source; this results in energy being scattered directly into the sensor field of view. This is the subject of the next section. Second, diffraction off the aperture and knife-edges results in energy being dispersed into the field of view. This section deals with diffraction and with imaging systems used to reduce this type of stray light.

14.4.1   Diffraction

Diffraction is a complex phenomenon and is difficult to model mathematically. The problem of diffraction in an optical baffle is best illustrated in terms of Fig. 14.7, which is a schematic representation of a baffle with a square entrance aperture. Radiant energy from the source s passes through the aperture and is incident upon the plane containing the point p.

In the absence of diffraction, the distribution of light in the plane is exactly that which is predicted by geometrical optics. A ray-trace analysis would show that for a point source at a distance $r_{10} = \infty$, the geometrical shape of the entrance aperture appears projected into the plane at P.

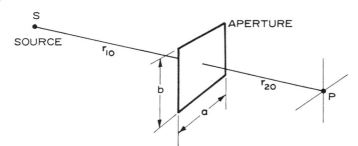

Figure 14.7 Schematic representation of a baffle with a square entrance aperture for diffracting on the entrance aperture into the plane at point P. (From J. M. Stone, *Radiation and Optics*, McGraw-Hill, New York, 1963, p. 193. Used with permission.)

However, the effect of diffraction is to cause fringes to appear near the edge of the pattern, and energy is spread into the shadow of the baffle.

According to Stone,[4] the effect of diffraction can be represented by the term Δv, which, by Fig. 14.7, is

$$\Delta v = a \left(\frac{2(r_{10} + r_{20})}{\lambda r_{10} r_{20}} \right)^{1/2} = \left(\frac{2a^2}{\lambda r_{20}} \right)^{1/2} \qquad (14.7)$$

where $r_{10} \gg r_{20}$ (point source at a distance).

Equation (14.7) shows that Δv depends upon the baffle width-to-wavelength ratio a/λ and the baffle width-to-length ratio a/r_{20}. The resulting diffraction patterns are shown in Fig. 14.8 as a function of Δv.

A quite complete idea of the nature of diffraction is obtained from Fig. 14.8. When Δv is greater than 10, the pattern of light obtained at the end of the baffle is roughly that predicted by geometrical optics, the principal diffraction being the prominent fringes near the edge of the pattern, which are referred to as *Fresnel patterns*.

The patterns obtained for small Δv are known as *Fraunhofer patterns*. In this case the distribution of energy is found to extend far into the geometrical shadow and is obtained directly as the Fourier transform of the aperture distribution.[5]

An objective in the design of the nth-order baffle is to use a length-to-width ratio sufficient to prevent direct illumination of the optical surfaces that are located at the end of the baffle tube. A baffle tube must be designed in the *Fresnel* region for two reasons: First, for Fraunhofer patterns (Δv less than unity), most of the on-axis energy is dispersed out of the optical path; hence the throughput is poor. Second, stray light from off-axis sources is diffracted into the field of view to a much greater extent for low Δv. A criterion for baffle design, based upon Fig. 14.8, is that $\Delta v \geq 50$ to avoid the problems of Fraunhofer diffraction. The following numerical example illustrates the use of Eq. (14.7) to place a limit on a baffle length-to-width ratio.

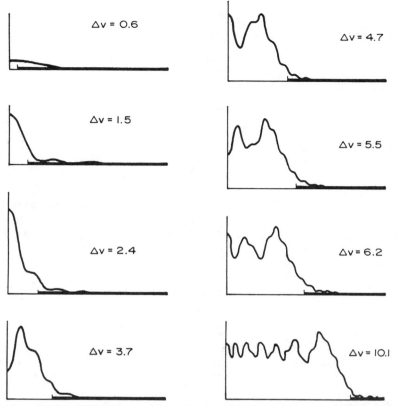

Figure 14.8 Diffraction patterns associated with a rectangular aperture for several values of Δv. The heavy segment along the horizontal axis indicates the region of the shadow of the entrance aperture. (From J. M. Stone, *Radiation and Optics*, McGraw-Hill, New York, 1963, p. 194. Used with permission.)

Example 1:   Find the maximum length of a baffle for an electro-optical sensor that has a 6-in. diameter collecting mirror and maximum wavelength of 30 μm.

Basic equation:

$$\Delta v = \left(\frac{2a^2}{\lambda r_{20}} \right)^{1/2} \tag{14.7}$$

Assumptions: The entrance aperture of the baffle is also 6 in.

Solution: Solve Eq. (14.7) for the baffle length r_{20}:

$$r_{20} = \frac{2a^2}{\lambda (\Delta v)^2} = \frac{2 \times (15.24[\text{cm}])^2}{30[\mu\text{m}] \times 10^{-4}[\text{cm}/\mu\text{m}] \times 50^2}$$

$$= 61.94 \text{ cm}$$

where $a = 6$ in., $\lambda = 30$ μm, and $\Delta v = 50$.

The length-to-width ratio is

$$L/W = 61.94/15.24 = 4.06$$

and the baffle angle is

$$\alpha = \arctan(1/4.06) = 13.82°$$

This problem illustrates that direct illumination of the collector optics cannot be avoided for off-axis sources less than approximately 13.82° off-axis for a 3-μm sensor. ∎

14.4.2 Optical Systems

The approach used to reduce stray light caused by diffraction is to establish a baffle entrance pupil as the limiting aperture, or entrance pupil, which is imaged somewhere in the system so the diffracted energy can be blocked with a glare stop.[6] This is represented through the use of thin-lens ray-tracing schematic of Fig. 14.9.

Diffraction at the entrance pupil P_1 is imaged at P_1', where it is blocked by the Lyot (glare) stop. The ray trace shows that energy diffracted in any direction from P_1 is imaged at P_1'. The "effective" aperture or entrance pupil P_2' is defined as the reverse image of P_2.

The glare stop reduces the throughput of the system by the ratio $(A_1/A_2)^2$, which is the penalty for reducing stray light originating near the optical axis. The loss depends upon the actual size of the glare stop.

The distance from P_1' to P_2 is determined by the quality of the image of P_1 at P_1'. The diffracted energy is spread over a finite region about P_1' because of the nonideal characteristics of the lenses L_1 and L_2.

The imaging system depicted in Fig. 14.9 accepts energy that is relatively collimated, as determined by the field stop, and outputs col-

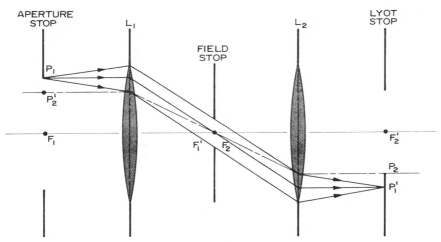

Figure 14.9 Optical schematic of an imaging system employing a glare stop to block energy diffracted from the entrance aperture.

limated energy. Generally, such systems include a condensing lens and a second field stop to collect the energy onto a detector. The basic scheme depicted in Fig. 14.9 has been employed in reflective systems.[7]

The actual performance of such systems can be predicted using various computer programs. Initially a "straight" Monte Carlo method was used. The physical shape of the system is modeled into the program. Individual rays are traced through the system. Whenever a ray intercepts a surface, the energy is decreased and then, depending upon the nature of the surface, is statistically scattered. The major limitation is that the number of rays that must be traced is proportional to the stray light transmittance. Thus, for systems capable of an off-axis rejection of 10^{-10}, only one ray in 10,000 would reach the detector. To obtain reliable statistics, many more rays must be traced.

Around 1970, the Space and Missiles System Organization (SAMSO) funded two projects to improve the approach. Honeywell, Inc. developed a "modified" Monte Carlo program called General Unwanted Energy Rejection Analysis Program, or GUERAP I. This program differed from the "straight" Monte Carlo program in that it permitted the "important" directions of scatter to be preferentially selected, allowing more rays to reach the detector. Each ray is weighted according to the modified statistics, so the results are the same but require less computer time.

Perkin-Elmer developed the second but deterministic approach, called GUERAP II. In this system, rays are propagated specularly from the entrance aperture to the detector. At each surface the energy is attenuated according to the properties of the material. In addition, diffuse scatter is calculated when the surface is "critical"; that is, it scatters directly onto the detector.

In 1972, NASA funded the development of another program: Arizona's Paraxial Analysis of Radiation Transfer (APART) at the Optical Science Center of the University of Arizona. This system is deterministic, like GUERAP II.

The software programs are not perfect. Modeling every mechanical detail of a baffle is extremely complex. The programs are difficult to implement and require experienced personnel. However, they are useful in trade-off studies, parametric analysis, and studies of alternative designs.[8]

14.4.3 BRDF

Low-scatter mirrors are evaluated for their surface quality in terms of a figure of merit referred to as the bidirectional reflectance distribution function (BRDF), which is defined as[9,10]

$$\text{BRDF}(\theta) = \frac{dL(\theta)}{dE(\theta)} \qquad [\text{sr}^{-1}] \qquad (14.8)$$

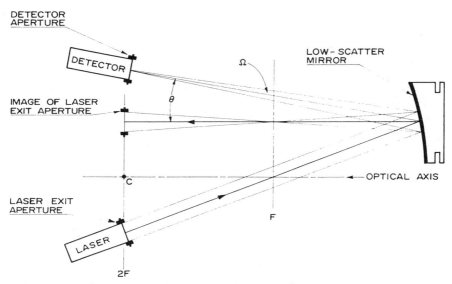

Figure 14.10 Geometry to measure BRDF(θ) for an off-axis parabolic mirror.

where L is the sterance [radiance] in W cm^{-2} sr^{-1}, E is the areance [irradiance] in W/cm^2, and θ is the off-axis angle.

The scattering of off-axis energy into the sensor field of view that occurs on the surface of a low-scatter mirror is a function of the off-axis angle.

The experimental setup for evaluating the BRDF[11] of an off-axis parabolic mirror is shown in Fig. 14.10. A collimated source (laser) is used to simulate an off-axis point source. This source is directed to the mirror. The imaged energy (on-axis) provides a measure of the power incident upon the mirror. The detector is moved off-axis to obtain the distribution of off-axis energy.

The detector entrance aperture is made equal in size to the image of the laser exit aperture, and the detector field of view matches the illuminated spot on the mirror. This assures that the detector measures all the power incident upon the mirror, Φ_m, for $\theta = 0$. The detector is then rotated to an off-axis angle θ to measure the scattered power $\Phi_s(\theta)$. The detector responds only to that part of the energy that is scattered into the solid angle Ω, defined by its aperture, which in this case is equal to its field of view. The BRDF is obtained as

$$\text{BRDF}(\theta) = \Phi_s(\theta)/\Phi_m\Omega \qquad (14.9)$$

where Φ_m is the total on-axis beam power, Φ_s is the power scattered at θ, and BRDF(θ) is the relative power scattered per unit solid angle at θ. When $\Phi_s = \Phi_m$ (on-axis), the BRDF has a value of $1/\Omega$.

For a linear detector (or for linearized data), Eq. (14.9) can be written as

$$\text{BRDF}(\theta) = V(\text{off axis})/V(\text{on axis})\Omega \qquad (14.10)$$

Equations (14.9) and (14.10) are equivalent to Eq. (14.8), since the voltage and/or power are related to the sterance [radiance] by a constant and since $E = L\Omega$.

Measurements of $\text{BRDF}(\theta)$ for flat optical surfaces require the use of imaging optics in the exit aperture[10] of the laser.

The ratio of off-axis to on-axis power for a point source is by definition the mirror off-axis response function $\mathcal{R}_m(\theta)$; thus, Eq. (14.9) can be written

$$\text{BRDF}(\theta) = \mathcal{R}_m(\theta)/\Omega \qquad (14.11)$$

The point-source off-axis mirror response function $\mathcal{R}_m(\theta)$ is obtained from Eq. (14.11) as

$$\mathcal{R}_m(\theta) = \text{BRDF}(\theta)\Omega \qquad (14.12)$$

where Ω is the sensor field of view. Equation (14.12) has a value of unity for $\theta = 0$ [see Eq. (14.9)].

14.5 EVALUATION OF THE OFF-AXIS RESPONSE

Figure 14.11 illustrates the geometry for the power incident upon the sensor aperture from an off-axis source. The differential throughput is given by

$$d\Upsilon = dA_s(\cos\theta)\frac{A_c}{s^2} = A_c\,d\Omega_c \qquad (14.13)$$

where the differential-source solid angle is given by

$$d\Omega_c = \cos\theta\frac{dA_s}{s^2} \qquad [\text{sr}] \qquad (14.14)$$

and where A_c is the effective aperture of the sensor and s is the distance.

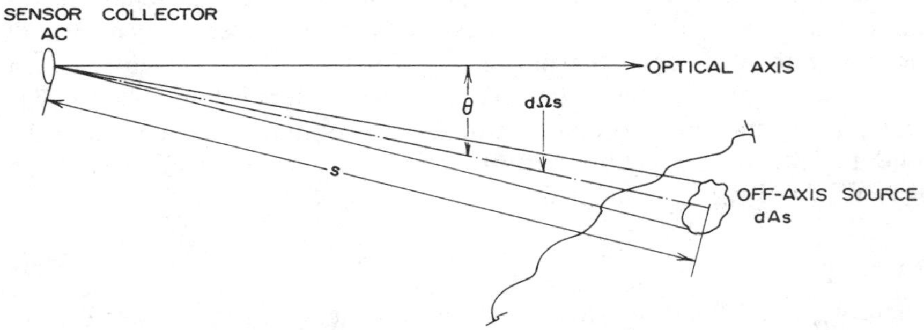

Figure 14.11 Illustration of the geometry for an extended-area off-axis source (the earth).

The weighted differential power on the aperture is given by

$$d\Phi_c = L_s \mathscr{R}(\theta) \, d\Upsilon = L_s \mathscr{R}(\theta) A_c \, d\Omega_c \quad [\text{W}] \qquad (14.15)$$

where L_s is the off-axis source sterance [radiance]. The effect of $\mathscr{R}(\theta)$ as a weighting factor in Eq. (14.15) is to yield the magnitude of the equivalent on-axis power that would evoke the same sensor response as does the off-axis source.

The total power incident upon the aperture is obtained by integration over the off-axis source.

$$\Phi_c = L_s A_c \int_{\text{off-axis source}} \mathscr{R}(\theta) \, d\Omega_c \qquad (14.16)$$

The equivalent on-axis sterance [radiance] is, by definition, the ratio of the power Φ_c to the sensor throughput:

$$L_c = \frac{\Phi_c}{A_c \Omega_c} = \frac{L_s}{\Omega_c} \int_{\text{off-axis source}} \mathscr{R}(\theta) \, d\Omega_c \qquad (14.17)$$

The off-axis response function $\mathscr{R}(\theta)$ in the above development includes the effects of low-scatter mirror BRDF and the forebaffle and is usually measured as part of the calibration.

The solution for Eq. (14.17) for a distant small-area source, such as the sun, is

$$L_c = L_s \mathscr{R}(\theta) \frac{\Omega_s}{\Omega_c} \qquad (14.18)$$

where θ is the sun angle (the angle between the sensor optical axis and the source) and Ω_s is the solid angle subtended by the sun at the sensor.

The solution for Eq. (14.17) for an extended-area off-axis source, such as the earth, requires numerical methods as illustrated in the following examples.

A prediction of the performance of a baffled low-scatter system is possible using system figures of merit, in a limiting case, as follows: With the assumption that the limiting stray light results from mirror scatter rather than from baffle knife-edge scatter, the effect of the baffle is to reduce the illuminated area of the low-scatter mirror. Then

$$\mathscr{R}(\theta) = \mathscr{R}_m(\theta) \mathscr{R}_b(\theta) \qquad (14.19)$$

where $\mathscr{R}_b(\theta)$ is the relative mirror area illuminated by the off-axis source. The relative area $\mathscr{R}_b(\theta)$ varies from unity for $\theta = 0$ to zero for the baffle angle α (see Fig. 14.12) and is obtained as the convolution of the entrance aperture with the mirror.

The following example is for an earth-limb sensor (Chap. 9) for which the sensor must view faint airglow emissions above the earth limb and reject the earth emissions, which constitute a relatively bright off-axis extended-area source.

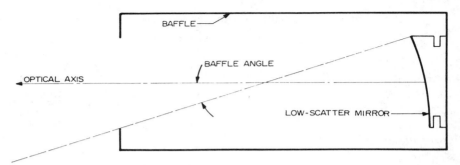

Figure 14.12 Physical layout of a baffled low-scatter mirror.

Example 2:   Estimate the quality of the off-axis mirror (BRDF) required to reduce the equivalent on-axis sterance from the earth as an off-axis source to a value equal to the system noise equivalent spectral sterance [radiance] (NESS). See Fig. 14.10.

Given: NESS = 3×10^{-9} W cm^{-2} sr^{-1} μm^{-1} at 10 μm. Baffle L/W = 3.49; α = arctan(1/3.49) = 16°; sensor height (above the earth) = 250 km. Off-axis angle (between earth and sensor optical axis) = 2.0°.

Basic equations:

$$\mathcal{R}_m(\theta) = \text{BRDF}(\theta)\Omega_c \qquad (14.12)$$

$$L_c = \frac{L_s}{\Omega_c} \int_{\text{earth}} \mathcal{R}(\theta)\, d\Omega_s \qquad (14.17)$$

$$\mathcal{R}(\theta) = \mathcal{R}_m(\theta)\mathcal{R}_b(\theta) \qquad (14.19)$$

Assumptions: (1) Stray light properties of the system are dominated by mirror scattering rather than baffle knife-edge scattering. (2) The relative area $\mathcal{R}_b(\theta)$ is obtained as the convolution of two circles of equal area—the mirror and the entrance aperture. (3) The earth radiates as a blackbody at 245 K.

Solution: An average value of BRDF can be obtained over the range of $\theta = 2$ to 16°. In this case (an extended off-axis source), most of the energy is incident upon the mirror at angles of 2 to 6°; therefore, the average BRDF is indicative of that required for such angles.
      Equation (14.17) can be written for L_c = NES and $\mathcal{R}_m(\theta)$ using Eq. (14.19) [where $\mathcal{R}_m(\theta)$ is the average value over $\theta = 2$ to 16°] as

$$\text{NES} = \frac{L_s}{\Omega_c}\mathcal{R}_m(\text{ave}) \int \mathcal{R}_b(\theta)\, d\Omega_s$$

and using Eq. (14.12),

$$\text{BRDF(ave)} = \mathcal{R}_m(\text{ave})/\Omega_c = \text{NES}\Big/ L_s \int \mathcal{R}_b(\theta)\, d\Omega_s$$

**Table 14.1   RELATIVE AREA FOR THE CONVOLUTION OF TWO
EQUAL-DIAMETER CIRCLES AND A BAFFLE ANGLE OF 16°**

Off-axis relative angle	Area
1	0.925370242221
2	0.849820383299
3	0.773942709144
4	0.698237638306
5	0.623145967968
6	0.549079099052
7	0.476445603617
8	0.405674959318
9	0.337240958904
10	0.271688593243
11	0.20967019296
12	0.152001326664
13	0.0997598404051
14	0.0544933046317
15	0.0187894462522
16	0.00

Table 14.1 gives the relative area $\mathscr{R}_b(\theta)$ for the convolution of two equal-diameter circles and a baffle angle of 16°. This tabulation is used in a computer program to calculate the integral of the earth at 250 km and 2° off axis. The result is

$$\int_{\text{earth}} \mathscr{R}_b(\theta) \, d\Omega_s = 0.026$$

which is given in Table 14.2. In addition, NESS is given as 3×10^{-9} W cm^{-2} sr^{-1} μm^{-1}, and the off-axis source (earth) sterance [radiance] $L_s(\lambda)$ is 3.36×10^{-4} W cm^{-2} sr^{-1} μm^{-1} (obtained from the solution to Planck's equation at 10 μm for 245 K; see Chap. 10). Thus,

$$\text{BRDF(ave)} = 3 \times 10^{-9}/(3.36 \times 10^{-4} \times 0.026) = 3.44 \times 10^{-4} \text{ sr}^{-1}$$

Note: BRDF(θ) values of 1×10^{-4} sr^{-1} at 1° off axis have been obtained in practical systems.   ■

**Table 14.2   WEIGHTED SOLID ANGLE OF THE EARTH AS A FUNCTION OF
SENSOR HEIGHT AND OFF-AXIS ANGLE FOR A 22° BAFFLE**

Altitude, km	Off-axis angle, deg		
	1	2	3
200	0.03005	0.02602	0.02213
250	0.02996	0.02598	0.02210
300	0.02994	0.02591	0.02202

The above example represents a realistic prediction of required BRDFs at angles greater than 2° off axis when the low-scatter mirror is directly illuminated.

The baffle term $\mathscr{R}_b(\theta)$ must be given a new interpretation for the case of a sunshade baffle in which the off-axis source angle exceeds the baffle angle. In this case the low-scatter mirror is never directly illuminated by the off-axis source. In such a system, the mirror functions as a second-order scatterer for which the total energy incident upon the baffle (the first-order scatterer) is scattered onto the mirror after having been attenuated by a fixed scattering constant [see Eq. (14.4)].

The following example illustrates the case of a sunshade baffle and low-scatter optics when direct illumination of the mirror does not occur.

Example 3:   Estimate the quality of the off-axis mirror (BRDF) required to reduce the equivalent on-axis sterance from the earth as an off-axis source to a value equal to the noise equivalent spectral sterance (NESS) for the baffle of Figs. 14.2 and 14.3.

Given: NESS $= 1 \times 10^{-11}$ W cm^{-2} sr^{-1} μm^{-1}; baffle angle $= 12°$ (see Fig. 14.3; earth angle $= 25°$).

Basic equations:

$$\mathscr{R}_m(\theta) = \text{BRDF}(\theta)\Omega_c \qquad (14.12)$$

$$L_c = \frac{L_s}{\Omega_c} \int \mathscr{R}(\theta)\, d\Omega_s \qquad (14.17)$$

$$\mathscr{R}(\theta) = \mathscr{R}_m(\theta)\mathscr{R}_b(\theta) \qquad (14.19)$$

Assumptions: (1) Primary mirror not illuminated since earth angle (25°) > baffle angle (12°). (2) The energy scattered by the baffle relative to the mirror is $\mathscr{R}_b(\theta) = 10^{-4}$, a constant (see Fig. 14.2). (3) Mirror BRDF$(\theta) = $ BRDF(ave), a constant.

Solution: An average value of BRDF can be obtained over the range of angles for which energy scatters onto the mirror from the baffle. $\mathscr{R}_b(\theta)$ is the energy scattered by the baffle relative to that scattered by the primary mirror $(3 \times 10^{-5} - 3 \times 10^{-9})$. Thus Eq. (14.19) is written for $\mathscr{R}_b(\theta) = 10^{-4}$. In addition, $L_c = $ NESS and $\mathscr{R}_m(\theta) = \mathscr{R}_m$(ave), so that using Eqs. (14.17) and (14.19),

$$\text{NESS} = \frac{L_s}{\Omega_c}\mathscr{R}_m(\text{ave})\mathscr{R}_b(\theta) \int d\Omega_s$$

and using Eq. (14.12),

$$\text{BRDF(ave)} = \frac{\mathscr{R}_m(\text{ave})}{\Omega_c} = \frac{\text{NESS}}{L_s\mathscr{R}_b(\theta) \int_{\text{earth}} d\Omega_s}$$

The term

$$\int_{earth} d\Omega_s = 0.827$$

which is obtained by integrating over the range of 22 to 90° off axis (over the earth). In addition,

$$NESS = 1 \times 1^{-11} \ W \ cm^{-2} \ sr^{-1} \ \mu m^{-1}$$

$$L_s(\lambda) = 3.36 \times 10^{-4} \ W \ cm^{-2} \ sr^{-1} \ \mu m^{-1} \ for \ T = 245 \ K$$

Thus

$$BRDF(ave) = \frac{1 \times 10^{-11}[sr^{-1}]}{3.36 \times 10^{-4} \times 1 \times 10^{-4} \times 0.827} = 3.60 \times 10^{-4} \ sr^{-1}$$

Note: In this case the baffle factor $\mathscr{R}_b(\theta) = 10^{-4}$ (a constant) is taken from the data of Fig. 14.2, which were measured in the visible. It is likely that $\mathscr{R}_b(\theta) < 10^{-4}$ for 10 μm; thus, this solution is conservative and the mirror BRDF obtained here should be a worst case.   ■

EXERCISES

1. Considering diffraction limits, find the minimum baffle angle for a sunshade baffle given that the clear aperture is 6 in. in diameter and operates at 24 μm. Use $\Delta v = 50$.

2. Given: $\lambda = 10 \ \mu$m, θ (off-axis angle) $= 30°$, $T(sun) = 5800$ K, $\Delta\theta(sun) = 0.53°$ (full-angle), $\mathscr{R}(\theta = 30°) = 10^{-10}$, $\Omega_c(sensor) = 10^{-3}$. Find the equivalent on-axis sterance [radiance] of the sun as an off-axis source for a telescoped radiometer. Assume the sun radiates as a blackbody, and neglect the effects of the intervening atmosphere.

3. Given: θ (off-axis angle) $= 10°$, sun parameters as in Exercise 2, $\Omega_c(sensor) = 10^{-4}$ sr, $\lambda = 10 \ \mu$m. Find the average BRDF of a mirror used in a radiometer telescope given that the equivalent on-axis sterance [radiance] is measured at $1 \times 10^{-9} \ W \ cm^{-2} \ sr^{-1}$ for the sun as an off-axis source. Assume: Direct illumination of the mirror through an equal area baffle (Table 14.1).

4. A low-scatter mirror has a bidirectional reflectance distribution function (BRDF) of $1 \times 10^{-4} \ sr^{-1}$ at 2° off axis. Given that the mirror is used as the primary element of a telescope for a radiometer with a solid-angle field of view of 1×10^{-3} sr, what is the point source off-axis response function at 2° for this system, assuming the mirror is fully illuminated?

5. Find the equivalent on-axis sterance [radiance] of the sun as an off-axis source for a telescoped radiometer, given that $\theta_s = 5°$ off-axis, $T(sun) = 5800$ K, mirror BRDF $= 3 \times 10^{-4} \ sr^{-1}$ at 5°, $\lambda = 10 \ \mu$m, and telescope field of view $\Omega_c = 10^{-5}$ sr. Assume that the sun radiates as a blackbody and that the mirror is fully illuminated by the sun, and neglect atmospheric losses. *Note:* The sun subtends an angle of 0.53° at earth.

REFERENCES

1. W. L. Wolf and G. J. Zissis, Eds., *The Infrared Handbook*, Office of Naval Research, Dept. of the Navy, Washington, DC, 1978, p. 8-100.
2. *Proc. Soc. Photo-Opt. Instrum. Eng.*, **107** (1977), entire volume concerning stray light problems in optical systems.
3. C. L. Wyatt, *Radiometric Calibration: Theory and Methods*, Academic Press, New York, 1978, pp. 3, 97, 119.
4. J. M. Stone, *Radiation and Optics*, McGraw-Hill, N.Y., 1963, pp. 190–193.
5. J. W. Goodman, *Introduction to Fourier Optics*, McGraw-Hill, N.Y., 1968, pp. 57–76.
6. P. J. Peters, "Aperture Shaping—A Technique for the Control of the Spatial Distribution of Diffracted Energy," *Proc. Soc. Photo-Opt. Instrum. Eng.*, **107**, 63–69 (1977).
7. R. M. Nadile et al., "SPIRE—Spectral Infrared Experiment," *Proc. Soc. Photo-Opt. Instr. Eng.*, **124**, 118–123 (1977).
8. R. P. Breault, "Problems and Techniques in Stray Radiation Suppression," *Proc. Soc. Photo-Opt. Instrum. Eng.*, **107**, 2–23 (1977).
9. R. J. Noll, "Surface Reflectance Models for Stray-light Calculations," *Proc. Soc. Photo-Opt. Instrum. Eng.*, **107**, 34 (1977).
10. F. E. Nicodemus, "Reflectance Nomenclature for Directional Reflectance and Emissivity," *Applied Optics*, **9**, 1474 (1970).
11. W. L. Wolfe, "Infrared BRDF Measurements," *Proc. Soc. Photo-Opt. Instrum. Eng.*, **107**, 173–180 (1977).

chapter 15

Thermal Detectors

15.1 INTRODUCTION

As implied by the name, thermal detectors respond to heat that is absorbed from the incident radiation and results in a change in the device temperature. Three of the most common thermal detectors are the bolometer, the thermocouple, and the pyroelectric detector.

In each of these detectors, the absorbed radiation produces a temperature change, which in turn alters a physical property of the material. The bolometer undergoes a change in electrical resistance, the thermocouple (or thermopile) produces a voltage at the junction of two dissimilar materials, and the pyroelectric detector undergoes a change in the polarization of the crystal.

The major advantages of these detectors are that (1) they respond uniformly to all wavelengths (for which the absorber has unity absorptance) and (2) they can be operated at room temperature. Their main disadvantages are relatively slow response time and lower sensitivity compared with photon detectors. An exception is the superconductive bolometer, which, when operated at about 1.5 K, exhibits exceptional performance.

The primary application of thermal detectors is as standard or reference detectors because of their uniform spectral response. For example, the electrically calibrated pyroelectric radiometer[1,2] provides a single-step-traceable standard to the measurement of electrical power. The thermal detector also has application in laboratory spectrometers used for measuring spectra over a broad wavelength range. Commercial detectors specify a usable spectral range from 0.2 to 35 μm depending primarily upon the window transmission and the receiver absorptivity.

15.2   THERMAL NOISE

Thermal detectors are sensitive to the rate of energy absorption that results in a rise in temperature. The fundamental limit to the detection of radiant energy in thermal detectors is understood in terms of the statistical nature of thermal equilibrium. The temperature of a body results from the random exchange of energy with its surroundings and fluctuates in a random fashion about its mean value. The purpose of this chapter is to set forth the basis for the thermal noise in the so-called ideal thermal detector.

15.2.1   Thermal Properties

When a temperature difference exists across a body, there results a flow of heat or energy. The thermal entities used to describe this heat flow are analogous to those describing the flow of electric current that results from a voltage difference in a circuit.

The thermal resistance R_T as defined by the temperature-energy relationship is

$$R_T = \frac{\Delta T}{\Delta \Phi} \quad [\text{K}/\text{W}] \tag{15.1}$$

where ΔT is the temperature differential and $\Delta \Phi$ is the incremental energy rate or power. The thermal conductance is $G_T = 1/R_T$. The thermal capacitance is defined by

$$C_T = \frac{\Delta U}{\Delta T} \quad [\text{J}/\text{K}] \tag{15.2}$$

where ΔU is the energy flow from a body resulting from a temperature difference ΔT.

From the above, the analogy with electric circuits is as follows

Thermal	Electrical
Temperature difference ΔT	Potential difference V
Energy rate Φ	Charge rate i
Thermal capacity C_T	Electrical capacity C
Thermal resistance R_T	Electrical resistance R

15.2.2   Thermal Fluctuations

A general theorem of statistical mechanics[3] gives the mean-square energy fluctuations as

$$\overline{\Delta U^2} = kT^2 C_T \quad [\text{J}^2] \tag{15.3}$$

for any system having many degrees of freedom. Based upon Eq. (15.2), the mean-square temperature fluctuation is

$$\overline{\Delta T_0^2} = \frac{kT^2}{C_T} \quad [\mathrm{K}^2] \tag{15.4}$$

It is now necessary to find the power spectrum; to do so, we resort to the electrical equivalent circuit for a radiation-coupled detector as illustrated in Fig. 15.1. The differential equation for the circuit of Fig. 15.1 is

$$C_T \frac{d(\Delta T)}{dt} + G_T \Delta T = \Delta \Phi \tag{15.5}$$

where $\Delta T = T_2 - T_1$ (Fig. 15.1). To solve this equation, let

$$\Delta \Phi = \Delta \Phi_0 \, e^{j\omega t} \tag{15.6}$$

and

$$\Delta T = \Delta T_0 \, e^{j\omega t} \tag{15.7}$$

where ΔT_0 and $\Delta \Phi_0$ are the mean incremental temperature and resulting flux, respectively. Then

$$\frac{d(\Delta T)}{dt} = j\omega \, \Delta T_0 \, e^{j\omega t} \tag{15.8}$$

and Eq. (15.5) becomes

$$C_T j\omega \, \Delta T_0 \, e^{j\omega t} + G_T \, \Delta T_0 \, e^{j\omega t} = \Delta \Phi_0 \, e^{j\omega t} \tag{15.9}$$

which gives

$$j C_T \omega \, \Delta T_0 + G_T \, \Delta T_0 = \Delta \Phi_0 \tag{15.10}$$

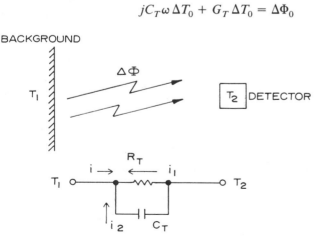

Figure 15.1 Top: Schematic representation of a detector coupled to its environment by radiation only. Bottom: Schematic circuit representation of the equivalent thermal circuit.

from which

$$\Delta T_0 = \frac{\Delta \Phi_0}{G_T + j\omega C_T} \quad [\mathrm{K}] \tag{15.11}$$

The mean-square temperature change is obtained by taking the absolute value (modulus) of ΔT_0

$$\overline{\Delta T_0^2} = \frac{\overline{\Delta \Phi_0^2}}{G_T^2 + \omega^2 C_T^2} \quad [\mathrm{K}^2] \tag{15.12}$$

It is of interest to know the frequency distribution of the temperature fluctuation. This is obtained by an appeal to Fourier analysis of random events. The temperature change results from a large number of very short impulses of energy. A Fourier analysis of a single impulse of this type is well known to show that the frequency spectrum is constant in amplitude up to very high frequencies. This is also true for the power spectrum of a series of random impulses, which therefore, on the average, is constant so that the power in a frequency interval is proportional to Δf. This result is in accordance with Rayleigh's theorem[4] that the mean-square values in neighboring frequency intervals are additive.

Thus, the mean-square power exchanged between the detector and the environment is given by

$$\overline{\Delta \Phi_0^2(f)} = B \Delta f \quad [\Phi/\mathrm{Hz}^{1/2}] \tag{15.13}$$

where B is constant. The variations in B at very high frequencies can be ignored, since the thermal inertia will not let the system respond to these high frequencies anyway.

Substitution of Eq. (15.13) into Eq. (15.12) gives

$$\overline{\Delta T_0^2(f)} = \frac{B \Delta f}{G_T^2 + \omega^2 C_T^2} \tag{15.14}$$

which when integrated over all frequencies yields

$$\overline{\Delta T_0^2} = \frac{B}{4 C_T G_T} \tag{15.15}$$

Eliminating $\overline{\Delta T_0^2}$ between Eqs. (15.15) and (15.4) yields

$$B = 4kT^2 G_T \tag{15.16}$$

and

$$\overline{\Delta T_0^2(f)} = \frac{4kT^2 G_T \Delta f}{G_T^2 + \omega^2 C_T^2} \tag{15.17}$$

For the case where the thermal capacity may be neglected, we have

$$\overline{\Delta T_0^2(f)} = 4kT^2R_T\,\Delta f \tag{15.18}$$

which is very similar to Eq. (7.19)—the Johnson noise fluctuations in a resistor—except that it contains T^2 rather than T. This results from the fact that according to the Carnot heat cycle only $\Delta T/T$ of ΔT is available for work, while in electrical circuits all the energy is available for work. It is also interesting that the power spectrum does not depend upon the thermal capacity C_T.

The power spectrum is obtained by combining Eqs. (15.16) and (15.13) to get

$$\overline{\Delta\Phi_0^2(f)} = 4kT^2G_T\,\Delta f \tag{15.19}$$

15.2.3 Thermal Conductance

In general, the value of thermal conductance G_T depends upon the way in which a body is thermally connected to its surroundings. We are interested in the simplest case for the ideal thermal detector—one in which the detector is connected to its surroundings by radiation coupling only. According to Stefan's law [Eq. (10.10)], the incremental heat flow from a body is given by

$$\Delta\Phi = 4A\sigma\varepsilon T^3\Delta T \tag{15.20}$$

when the temperature difference ΔT is small, where A is the surface area and ε is the emissivity of the body. Using Eq. (15.1) the thermal conductance for a radiation coupled detector is found to be

$$G_T = 4A\sigma\varepsilon T^3 \tag{15.21}$$

15.2.4 Noise Equivalent Power (NEP)

Substitution of Eq. (15.21) into (15.19) yields the mean-square power fluctuations for the ideal thermal detector:

$$\overline{\Delta\Phi_0^2} = 16kT^5\sigma\varepsilon A\,\Delta f \quad [\text{W}^2] \tag{15.22}$$

Example 1: Determine the NEP and D^* for the ideal thermal detector.

Given: The detector parameters are $A = 0.01$ cm^2, $T = 300$ K, $\varepsilon = 1$, and $\Delta f = 1$ Hz.

Basic equations:

$$\overline{\Delta\Phi_0^2} = 16kT^5\sigma\varepsilon A\,\Delta f \quad [\text{W}^2] \tag{15.22}$$

$$D^* = (A_d\,\Delta f)^{1/2}/\text{NEP} \quad [\text{cm Hz}^{1/2}/\text{W}] \tag{6.3}$$

Assumptions: The detector is coupled to the environment by radiation only, and all other noise forms are negligible.

Solution:

$$\overline{\Delta\Phi_0^2} = 16 \times 1.38 \times 10^{-23}[\text{J K}^1] \times 300^5[\text{K}^5] \times 5.669 \times 10^{-8}[\text{W m}^{-2}\text{ K}^{-4}]$$
$$\times 1 \times 10^{-2}[\text{cm}^2] \times 1[\text{Hz}] \times 10^{-4}[\text{m}^2/\text{cm}^2]$$
$$= 3.042 \times 10^{-23}\text{ W}^2$$
$$\text{NEP} = \left(\overline{\Delta\Phi_0^2}\right)^{1/2} = 5.515 \times 10^{-12}\text{ W}$$

and

$$D^* = (0.01 \times 1)^{1/2}/(5.515 \times 10^{-12}) = 1.8 \times 10^{10}\text{ cm Hz}^{1/2}/\text{W}$$

Practical detectors are designed to approximate the ideal detector through (1) mounting the sensitive element within a vacuum to eliminate convective coupling and (2) utilizing very small electrical leads to reduce conductive coupling.

A typical D^* value is 2×10^8 cm Hz$^{1/2}$/W.   ∎

15.3 BOLOMETER

The bolometer is essentially a resistor that exhibits a relatively large temperature coefficient of resistance. The sensitive element typically consists of a thin film or flake of semiconductor material. Thermistors exhibit similar properties. The infrared bolometer detector is coated with optically black material to increase its absorption.

The change in electrical resistance on heating depends upon a quantity α, known as the temperature coefficient of resistance, which is defined[5] by

$$\alpha = \frac{1}{R_d}\frac{dR_d}{dT} \qquad [\text{K}^{-1}] \tag{15.23}$$

where R_d is the detector resistance. For metals, α has values ranging from 0.003 to 0.006 K^{-1} at room temperature, and for semiconductors α is an order of magnitude larger.

The response time and responsivity depend upon the rate of heat transfer between the active element and its heat sink. Greater thermal contact results in faster response but reduces the magnitude of the temperature change that is possible. Hence, response time and responsivity must be traded off. The responsivity for an ideal metal bolometer with predominantly conductive cooling is of the order of 30 to 80 V/W, while that of a semiconductor bolometer[6] is of the order of 10^4 V/W.

The noise mechanism in thermistor bolometers includes current noise resulting from bias current and thermal noise. Current noise is also referred to as $1/f$, excess, or modulation noise.

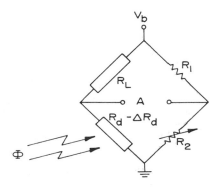

Figure 15.2 Use of matched flakes for which slow changes in ambient conditions cancel. The bridge circuit cancels the effect of dc bias.

Typical performance[7] for room-temperature bolometers is $D^* \approx 1 \times 10^6$ cm Hz$^{1/2}$/W, and the time constant $T_c = 1$ ms.

There are two detector configurations available. One is used for dc measurements and the other for chopped radiation. Detectors for dc measurements consist of two matched detector flakes; the active one is coated with absorbant material and the other, a "dummy" detector, is shielded from radiation. This configuration is used in a bridge circuit as shown in Fig. 15.2. Slow changes in ambient temperature affect each element equally, thus canceling out any effect.

The change in the voltage at A (Fig. 15.2) due to a change ΔR_d, provided $\Delta R_d \ll R_d$, is

$$\Delta V = \frac{R_L V_b \Delta R_d}{(R_d + R_L)^2} = \frac{R_L i \Delta R_d}{R_d + R_L} \qquad [\text{V}] \qquad (15.24)$$

where V_b is the bias voltage, R_L is the resistance of the dummy detector, and ΔR_d is the change in the detector resistance in response to incident flux.

For chopped sources, the second half of the bridge is replaced with a capacitor to block the dc voltage, and R_L can be any suitable load resistor. In this configuration, only a single detector element is used. Equation (15.24) still applies in this case, and ΔV is the peak-to-peak voltage resulting from the resistance change ΔR_d.

15.4 THERMOCOUPLES AND THERMOPILES

A thermocouple consists of a junction of two dissimilar materials that when heated produced a voltage across the two open leads. This is the thermovoltaic effect. When more than one junction is combined into a responsive element, it is termed a thermopile.

Thermopiles are fabricated using thin-film techniques that permit high-density junctions and complex detector arrays. There exists a trade-off between responsivity and time constant as with the bolometer. The main advantage of the thermopile is that a bias current is not required, which is convenient; in addition, it eliminates the current noise so that thermal noise predominates.

The Seebeck effect[8] relates to the electromotive force (EMF) produced in a circuit made up of different conducting elements when the contacts are at different temperatures. The open-circuit voltage is given by the product of the thermoelectric power Φ_{AB} and the temperature differential ΔT.

$$V_o = \Phi_{AB}\,\Delta T \qquad (15.25)$$

The quantity Φ_{AB} is characteristic of the two materials. For example, a junction consisting of antimony and bismuth exhibits a thermoelectric power of 100 μV/C. If current is allowed to flow in the circuit, the Peltier effect[8] reduces the responsivity. The Peltier effect refers to the cooling of the junction when current flows through it.

Responsivities for metal thermocouples vary from 0.4 to 0.6 V/W, with time constants roughly equal to 10 ms. The NEP is about 2×10^{-10} W/Hz$^{1/2}$.

Semiconductor thermopiles exhibit higher thermoelectric power than pure metals; however, the resistance and noise are higher also. Responsivities in the order of 10 to 50 V/W are available with $D^* \approx 1 \times 10^8$ cm Hz$^{1/2}$/W.

The frequency response of the thermopile favors low frequency and dc applications that take advantage of the detector's inherent dc stability.

15.5 PYROELECTRIC DETECTORS

Pyroelectric detectors share many characteristics with other thermal detectors. In general they are capable of uniform response over a broad wavelength range, limited by the ability of the receiver to absorb radiation.

Unlike other thermal detectors, the pyroelectric effect depends upon the rate of change of the detector temperature, rather than the absolute temperature. This results in a much faster time constant than those of thermopiles or bolometers. It also means that pyroelectric detectors respond only to chopped or pulsed radiation and the steady background radiation has no effect.

The physical effect of pyroelectric detection is the temperature dependence of the electric polarization of certain materials of which ferroelectric crystals are the most common. These crystals exhibit an internal electric field along a certain crystal axis. The electric field results from the alignment of electric dipole moments. The electric field is directly proportional to crystal temperature, because the degree of alignment (or "polarization") can be disturbed by photon vibration. When the crystal is an insulating material and parallel electrodes are attached, the electric charge is attracted and stored on the plates. Thus, if the temperature changes, the electric field varies also. This change produces an observable current when the electrodes are connected to external circuits.

A simple detector consists of the pyroelectric crystal coated on its front and back with electrodes and mounted in an enclosure that often includes an FET amplifier and matched load resistor.

When a very-low-load (50-ohm) resistor is shunted across the detector, an effective response time of nanoseconds can be attained. However, the responsivity is reduced dramatically. Typically a 10^8-ohm load yields a time constant of 1 ms with $D^* \approx 1 \times 10^8$ cm $Hz^{1/2}/W$.

EXERCISES

1. Assume constant bias voltage for the bridge circuit of Fig. 15.2 and Eq. (15.24). What is the optimum value of R_L for maximum output voltage ΔV?
2. Assume constant bias current for the bridge circuit of Fig. 15.2 and Eq. (15.24). What is the optimum value of R_L for maximum output voltage ΔV?
3. Find NEP and D^* for an ideal thermal detector of area 0.05 cm^2 at a temperature of 193 K (Dry Ice) and noise bandwidth of 1 Hz. Assume the detector is coupled to the environment by radiation only and its receiver has an $\varepsilon = 1$.
4. A pyroelectric detector exhibits maximum D^* at a chopping frequency of 10 Hz. The optical signal is detected and dc-restored using a phase-sensitive rectifier and low-pass filter. What is the maximum rise time of the incoming flux permitted to avoid aliasing errors? What is the low-pass filter time constant?
5. A 1-mm diameter thermopile detector has a D^* of 2×10^8 cm $Hz^{1/2}/W$. If it is used in a system with an information/noise bandwidth of 5 Hz, what is the detector NEP?
6. For the detector bridge circuit of Fig. 15.2, given $R_d = R_L = 5 \times 10^7$ ohms, $V_b = 100$ V, and $\Delta R_d = 1 \times 10^{-4}$ ohms for $\Delta \Phi = 1 \times 10^{-12}$ W incident upon the detector, find the detector responsivity in units of volts per watt.

REFERENCES

1. Jon Geist et al., "Electrically Based Spectral Power Measurements through Use of a Tunable cw Laser," *Applied Physics Letters*, **26**, 309–311 (1975).
2. R. J. Phelan, Jr., and A. R. Cook, "Electrically Calibrated Pyroelectric Optical-Radiation Detector," *Applied Optics*, **12**, 2494–2500 (1973).
3. R. A. Smith, F. E. Jones, and R. P. Chasmar, *The Detection and Measurement of Infrared Radiation*, Oxford University Press, London, 1957, p. 205.
4. Ibid., p. 176.
5. Ibid., p. 94.
6. T. Limperis and J. Mudar, "Detectors," in W. F. Wolfe and G. J. Zissis, Eds., *The Infrared Handbook*, Office of Naval Research, Washington, DC, 1978, Ch. 11, pp. 11–23.
7. Editors, "The Optical Industries & Systems Encyclopedia & Dictionary," *Optical Spectra*, 1979, Vol. 2, p. E52.
8. E. V. Condon and H. Odishaw, Eds., *Handbook of Physics*, McGraw-Hill, N.Y., 1958, p. 4-84.

chapter 16

Photon Noise
Limited Detectors

16.1 INTRODUCTION

Photon detectors are sensitive to the rate of photon absorption, whereas thermal detectors respond to energy rate or power. *Photon noise* is the term describing fluctuations in the instantaneous value of the number of photons in a beam of radiation. These fluctuations set a fundamental limit to the accuracy by which the average photon rate can be measured that arises from the quantum nature of the radiant source. This limitation applies to the case of an ideal detector that is capable of counting every incident photon and generates no internal noise.

Infrared photon detectors are often operated at low temperatures of the order of 80 K or lower. Generally they are surrounded by a cold enclosure at the same temperature. Radiation can enter the enclosure through an aperture; thus, the detector is irradiated by flux originating from the target, the background, and even the window.

Uncoded radiation reaching the detector from its surroundings, or from the target background, produces a fixed photon noise level that sets a limit to the level that can be detected from the target. This condition is referred to as a BLIP (background-limited infrared photoconductor).

Photon noise, in photon-counting detectors, is similar to shot noise in a vacuum tube. Unlike shot noise, it does not depend upon the fixed value of the electronic charge e but rather on the quantum energy associated with the photon, which can have any value, depending upon the frequency of radiation.

In photoemissive, photovoltaic, or photoelectromagnetic detectors, only photons contribute to the noise. But photoconductive detector noise depends upon the change in the concentration of charge carriers. The concentration of charge carriers is determined by both the generation and recombination rates. Fluctuations in these rates are termed generation-recombination (gr) noise. It has been shown[1] that at equilibrium in a photoconductor, the total noise, photon and gr, can be no less than twice the photon noise alone. Since D^* depends upon the square root of the noise power, the photon noise limit of photoconductors is $\sqrt{2}$ times poorer than for photovoltaic detectors.

16.2 FLUCTUATIONS IN BLACKBODY FLUX

The fluctuations in the photon sterance [radiance] of a beam of blackbody radiation are given in this section. Planck's equation for spectral radiant sterance [radiance] is

$$L_e(\lambda) = \frac{2hc^2}{\lambda^5} \frac{1}{\exp(hc/\lambda kT) - 1} \qquad [\text{W m}^{-3} \text{ sr}^{-1}] \qquad (16.1)$$

where h is Planck's constant in J s; c is the velocity of light, in m/s, k is Boltzmann's constant, in J/K; λ is the wavelength in m, and T is the temperature in degrees kelvin.

The energy in a photon is $E_p = h\nu = hc/\lambda$, where ν is the optical frequency in Hz. The relationship between radiant flux Φ_e and photon flux Φ_p is therefore

$$\Phi_e = \Phi_p [\text{q/s}] \frac{h[\text{J s}]c[\text{m/s}]}{\lambda[\text{m}]} \qquad [\text{J/s}] \qquad (16.2)$$

and the derivative of Eq. (16.2) yields

$$\frac{d\Phi_p}{d\Phi_e} = \frac{\lambda}{hc} \qquad \text{or} \qquad 1 = \frac{d\Phi_p hc}{d\Phi_e \lambda} \qquad (16.3)$$

Using the definition of sterance, [Eq. (3.19)] and Eqs. (16.1) and (16.3), the photon sterance is found to be

$$L_p = \frac{d\Phi_p}{d\Upsilon} \frac{d\Phi_e}{d\Phi_p} \frac{\lambda}{hc} = L_e \frac{\lambda}{hc} \qquad [\text{q s}^{-1} \text{ m}^{-2} \text{ sr}^{-1}] \qquad (16.4)$$

Thus,

$$L_p(\lambda) = \frac{2c}{\lambda^4} \frac{1}{\exp(hc/\lambda kT) - 1} \qquad (16.5)$$

where $L_p(\lambda)$ can be thought of as the average spectral photon sterance in a beam.

According to the Bose-Einstein relation,[2] the mean-square fluctuation in the photon sterance for discrete events is given by

$$\overline{\Delta L_p^2(\lambda)} = L_p(\lambda) \frac{\exp(hc/\lambda kT)}{\exp(hc/\lambda kT) - 1} \tag{16.6}$$

Thus, the mean-square fluctuation in the photon sterance for a blackbody source is

$$\overline{\Delta L_p^2(\lambda)} = \frac{2c}{\lambda^4} \frac{1}{\exp(hc/\lambda kT) - 1} \frac{\exp(hc/\lambda kT)}{\exp(hc/\lambda kT) - 1} \tag{16.7}$$

The photon detectors of interest are generally sensitive to wavelengths < 30 μm and are often operated at reduced temperatures for which $hc/\lambda \gg kT$, which means that $\exp(hc/\lambda kT) \gg 1$. Representative values of the exponent are given in the following tabulation:

Wavelength, μm	Temperature, K			
	5	80	193	300
1	∞	5.4×10^{77}	1.5×10^{32}	5.2×10^{20}
10	∞	5.9×10^7	1.65×10^3	1.2×10^2
100	2.9×10^{12}	5.9	2.10	1.61

This permits the following approximation for the mean-square fluctuations:

$$\overline{\Delta L_p^2(\lambda)} = L_p(\lambda) = \frac{2c}{\lambda^4} \exp\left(\frac{-hc}{\lambda kT}\right) \tag{16.8}$$

which is interpreted as meaning that the mean-square fluctuations in the sterance are approximately equal to the mean sterance.

The total mean-square fluctuation in the mean photon sterance is given by integrating Eq. (16.8) over the appropriate wavelengths:

$$\overline{\Delta L_p^2} = L_p = 2c \int_{\lambda_1}^{\lambda_2} \lambda^{-4} \exp\left(\frac{-hc}{\lambda kT}\right) d\lambda \tag{16.9}$$

An analytical solution to Eq. (16.9) for $\lambda_1 = 0$ and $\lambda_2 = \lambda_c$ (the detector cutoff wavelength) is given by

$$\overline{\Delta L_p^2} = L_p = 2c \frac{kT}{hc} \left[2\left(\frac{kT}{hc}\right)^2 + \frac{2kT}{hc\lambda_c} + \frac{1}{\lambda_c^2} \right] \exp\left(\frac{-hc}{\lambda_c kT}\right) \tag{16.10}$$

which applies to unfiltered blackbody radiation.

Another case of interest is when the photons are limited by a spectral bandpass filter. In this case it is convenient to use numerical techniques to integrate Eq. (16.9) to obtain L_p as follows.

$$\overline{\Delta L_p^2} = L_p = 2c \Delta\lambda \sum_{i=1}^{n} \lambda_i^{-4} \mathscr{R}_i \exp\left(\frac{-hc}{\lambda_i kT}\right) \tag{16.11}$$

where $\Delta\lambda$ is the incremental wavelength and $\mathcal{R}_i$ is the normalized filter transmittance at the wavelength λ_i. The summation is over all nonzero values of $\mathcal{R}_i$.

Equation (16.11) can often be approximated assuming the equivalent ideal filter as

$$\overline{\Delta L_p^2} = L_p = 2c(\lambda_2 - \lambda_1)\lambda_0^{-4}\exp\left(\frac{-hc}{\lambda_0 kT}\right) \qquad (16.12)$$

where $\lambda_2 - \lambda_1$ is the equivalent ideal bandwidth and λ_0 is the center wavelength.

16.3 PHOTON NOISE

The photon noise in detectors arises from fluctuations in the *arrival* rate of incident photons. Such detectors are often operated within a cold shield as illustrated in Fig. 16.1. The mean-square fluctuation $\overline{\Delta\Phi_p^2}$ in the background flux incident upon the detector is obtained using Eq. (3.21) and in accordance with the approximation given in Eq. (16.8):

$$\overline{\Delta\Phi_p^2} = L_p\Upsilon' \qquad [\text{q/s}] \qquad (16.13)$$

where the mean-square fluctuation in the sterance is given in terms of the average sterance. Equation (16.13) is expressed for the background radiation incident upon the detector, where Υ' is the detector throughput and L_p is the background sterance incident upon the detector entrance aperture.

The mean-square charge carrier generation rate noise is similar to shot noise:

$$\overline{G_n^2} = 2\overline{\Delta\Phi_p^2}Q\,\Delta f = 2\,\Delta f Q L_p\Upsilon' \qquad [\text{C}^2/\text{s}^2] \qquad (16.14)$$

(where C is the unit of charge in coulombs) except that it depends upon the mean flux rather than the mean current. The quantum efficiency Q is the

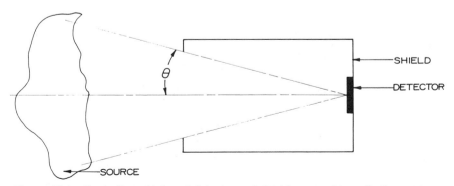

Figure 16.1 Illustration of infrared detector and shield exposed to radiant source.

number of charge carriers produced per photon incident upon the detector. The rms noise charge carrier generation rate is given by

$$G_n(\text{rms}) = \left(2\,\Delta f Q L_p \Upsilon'\right)^{1/2} \quad [\text{C/s}] \tag{16.15}$$

The signal charge carrier generation rate is given by

$$G_s = \Phi_p Q = \Phi_e Q \lambda / hc \quad [\text{C/s}] \tag{16.16}$$

for monochromatic radiation at λ, where Φ_e is the incident flux in watts. The responsivity is obtained from Eq. (16.16) as

$$\mathcal{R}_c = G_s / \Phi_e = Q\lambda / hc \quad [\text{C s}^{-1}\,\text{W}^{-1}] \tag{16.17}$$

and the detector NEP is given by

$$\text{NEP}_\lambda = \frac{G_n}{\mathcal{R}_c} = \frac{hc}{\lambda}\left(\frac{2\,\Delta f L_p \Upsilon'}{Q}\right)^{1/2} \quad [\text{W}] \tag{16.18}$$

where L_p is obtained by Eq. (16.9) appropriately integrated as in Eq. (16.10), (16.11), or (16.12). Substitution of L_p from Eq. (16.10) into Eq. (16.18) yields the NEP for the case of Fig. 16.1 as

$$\text{NEP}(\lambda) = \frac{2c}{\lambda}\left(\frac{kTh\,\Delta f\,\Upsilon'}{Q}\right)^{1/2}\left[2\left(\frac{kT}{hc}\right)^2 + \frac{2kT}{hc\lambda_c} + \frac{1}{\lambda_c^2}\right]^{1/2}\exp\left(\frac{-hc}{2kT\lambda_c}\right) \tag{16.19}$$

and the corresponding detectivity is

$$
\begin{aligned}
D^* &= \frac{(A\,\Delta f)^{1/2}}{\text{NEP}(\lambda)} \\
&= \frac{\lambda}{2c}\left(\frac{Q}{kTh\pi}\right)^{1/2}\exp\left(\frac{hc}{2kT\lambda_c}\right)\left[2\left(\frac{kT}{hc}\right)^2 + \frac{2kT}{hc\lambda_c} + \frac{1}{\lambda_c^2}\right]^{-1/2}(\sin\theta)^{-1} \\
&\quad [\text{m Hz}^{1/2}/\text{W}]
\end{aligned}
\tag{16.20}
$$

where the detector throughput $\Upsilon' = A\pi\sin^2\theta$.

The application of the above to the prediction of *system* performance in the case of the noise equivalent sterance [radiance] (NES) is obtained by using Eq. (16.18) and the definition of sterance [radiance] to obtain

$$\text{NES} = \text{NEP}/\Upsilon\tau_e \quad [\text{W cm}^{-2}\,\text{sr}^{-1}] \tag{16.21}$$

where Υ is the system throughput and τ_e is the system optical efficiency (including chopping and rectification bandwidth factors where appropriate).

The first case of interest corresponds to Eq. (16.10), where the detector is exposed to unfiltered background radiation with a throughput Υ' as in Fig. 16.1 but with the effective *system* throughput $\Upsilon\tau_e$ as in Eq. (16.21). The NES is obtained by substituting Eq. (16.19) into Eq. (16.21) to get

$$\text{NES}_\lambda = \frac{2c}{\lambda \Upsilon \tau_e} \left(\frac{k Th \Upsilon' \Delta f}{Q} \right)^{1/2} \left[2 \left(\frac{kT}{hc} \right)^2 + \frac{2kT}{hc\lambda_c} + \frac{1}{\lambda_c^2} \right]^{1/2} \exp\left(\frac{-hc}{2kT\lambda_c} \right)$$
(16.22)

The second case of interest is where the aperture of Fig. 16.1 is filled with a cold bandpass filter and the NES is obtained by substituting Eq. (16.11) or (16.12) into Eq. (16.21) for L_p.

A special case occurs when the *only* flux reaching the detector is coded "signal." This happens, for example, in cryogenic sensors that have the entire optical subsection cooled to a level for which the background is negligible.[3-5] In this case the detector throughput is identical with the effective system throughput, and Eq. (16.22) is written as

$$\text{NES}_\lambda = \frac{2c}{\lambda \tau_e} \left(\frac{k Th \, \Delta f}{Q \Upsilon} \right)^{1/2} \left[2 \left(\frac{kT}{hc} \right)^2 + \frac{2kT}{hc\lambda_c} + \frac{1}{\lambda_c^2} \right]^{1/2} \exp\left(\frac{-hc}{2kT\lambda_c} \right)$$
(16.23)

The resulting noise is proportional to the signal. This form of photon noise may be termed "noise in signal" and is significant in multiplexed spectrometers.[6]

An interferometer is a multiplex spectrometer that simultaneously samples and encodes the flux at all wavelengths within the free spectral range; thus, the photons at all wavelengths contribute to the noise. The following examples illustrate photon noise limitations in an interferometer.

Example 1:   Find the noise equivalent sterance [radiance] (NES) at 3.0 μm for a wide-band cryogenic interferometer irradiated with 300 K blackbody background.

Given: The interferometer has a free spectral range of 2 to 24 μm, the system throughput is 2.41×10^3 cm^2 sr, the optical efficiency over the free spectral range is constant at 0.19. The detector quantum efficiency is 0.5 between 8 and 30 μm, and the electrical bandwidth Δf is 2600 Hz.

Basic equation:

$$\text{NES}_\lambda = \frac{2c}{\lambda \tau_e} \left(\frac{k Th \, \Delta f}{Q \Upsilon} \right)^{1/2} \left[2 \left(\frac{kT}{hc} \right)^2 + \frac{2kT}{hc\lambda_c} + \frac{1}{\lambda_c^2} \right]^{1/2} \exp\left(\frac{-hc}{2kT\lambda_c} \right)$$
(16.23)

Assumptions: The integration implied in the equation for NES_λ, from 0 to λ_c, is a valid approximation where the cutoff wavelength λ_c is 24 μm, and the quantum efficiency is a constant because most of the energy from a 300-K blackbody is emitted above 8 μm, where the detector performance is nominal.

Solution:

$$\lambda = 3 \times 10^{-6} \text{ m}$$

$$\lambda_c = 24 \times 10^{-6}$$

$$2c = 2 \times 3 \times 10^8 \text{ m/s}$$

$$\frac{kT}{hc} = \frac{1.38 \times 10^{-23}[\text{J/K}] \times 300[\text{K}]}{6.6 \times 10^{-34}[\text{J s}] \times 3 \times 10^8[\text{m/s}]} = 2.09 \times 10^4 \text{ m}^{-1}$$

$$\left(\frac{kT}{hc}\right)^2 = 4.37 \times 10^8 \text{ m}^{-2}$$

$$\frac{2kT}{hc\lambda_c} = \frac{2 \times 2.09 \times 10^4[\text{m}^{-1}]}{24 \times 10^{-6}[\text{m}]} = 1.74 \times 10^9 \text{ m}^{-2}$$

$$\frac{1}{\lambda_c^2} = \left(\frac{1}{24 \times 10^{-6}[\text{m}]}\right)^2 = 1.74 \times 10^9 \text{ m}^{-2}$$

The term inside the square brackets in Eq. (16.23) is

$$[\ ] = 2 \times 4.37 \times 10^8[\text{m}^{-2}] + 1.74 \times 10^9[\text{m}^{-2}] + 1.74 \times 10^9[\text{m}^{-2}]$$

$$= 4.35 \times 10^9 \text{ m}^{-2}$$

$$[\ ]^{1/2} = 6.60 \times 10^4 \text{ m}^{-1}$$

$$\frac{kTh\,\Delta f}{Q\Upsilon} =$$

$$= \frac{1.38 \times 10^{-23}[\text{J/K}] \times 300[\text{K}] \times 6.6 \times 10^{-34}[\text{J s}] \times 2.6 \times 10^3[\text{s}^{-1}]}{0.5 \times 2.4 \times 10^{-3}[\text{cm}^2 \text{ sr}] \times 1 \times 10^{-4}[\text{m}^2/\text{cm}^2]}$$

$$= 5.93 \times 10^{-44} \text{ J}^2/\text{m}^2$$

$$\left(\frac{kTh\,\Delta f}{Q\Upsilon}\right)^{1/2} = 2.43 \times 10^{-22} \text{ J/m}$$

$$\frac{2c}{\lambda\tau} = \frac{2 \times 3 \times 10^8[\text{m/s}]}{3 \times 10^{-6}[\text{m}] \times 0.19} = 1.05 \times 10^{15} \text{ s}^{-1}$$

$$\frac{-hc}{2kT\lambda_c} = \frac{-1}{2 \times 2.09 \times 10^4[1/\text{m}] \times 24 \times 10^{-6}[\text{m}]}$$

$$= -1.00[\text{unitless}]$$

$$e^{-1} = 3.68 \times 10^{-1}$$

$$\text{NES}|_{3.0\,\mu m} = 1.05 \times 10^{15}[\text{s}^{-1}] \times 2.43 \times 10^{-22}[\text{J/m}]$$

$$\times 6.60 \times 10^4[\text{m}^{-1}] \times 3.68 \times 10^{-1}$$

$$= 6.20 \times 10^{-3} \text{ W m}^{-2} \text{ sr}^{-1} = 6.20 \times 10^{-7} \text{ W cm}^{-2} \text{ sr}^{-1}$$

Note: This problem illustrates the effect of wide-band background flux upon the photon noise limited sterance [radiance]. ∎

In the above example, the 300 K blackbody radiation is small below 8 μm. Thus, it is technically possible to measure faint signals at 3.0 μm in the presence of the 300 K radiation provided the photon noise is not excessive.

A common practice is to use a cold bandpass filter to limit the background. For example, a cold (nonemitting) filter centered at 3.0 μm with a 2-μm bandwidth substantially reduces the NES. The following example illustrates this practical method.

Example 2: Find the noise equivalent sterance [radiance] at 3.0 μm for a narrow-band cryogenic interferometer irradiated with 300 K blackbody background.

Given: The interferometer free spectral range is limited by a cold filter between 2 and 4 μm. The other important parameters correspond to those of Example 1.

Basic equations:

$$\overline{\Delta L_p^2} = L_p = 2c(\lambda_2 - \lambda_1)\lambda_0^{-4}\exp(-hc/\lambda_0 kT) \quad (16.12)$$

$$NEP_\lambda = \frac{hc}{\lambda}\left(\frac{2\,\Delta f L_p \Upsilon'}{Q}\right)^{1/2} \quad (16.18)$$

$$NES = NEP/\Upsilon\tau_e \quad [\text{W cm}^{-2}\,\text{sr}^{-1}] \quad (16.21)$$

Solution: Note that $\Upsilon' = \Upsilon$; thus

$$NES_\lambda = \frac{hc}{\lambda\tau_e}\left(\frac{2\,\Delta f L_p}{Q\Upsilon}\right)^{1/2}$$

$$\lambda_2 - \lambda_1 = 4 \times 10^{-6}\,\text{m} - 2 \times 10^{-6}\,\text{m} = 2 \times 10^{-6}\,\text{m}$$

$$2c = 2 \times 3 \times 10^8[\text{m/s}] = 6 \times 10^8\,\text{m/s}$$

$$\frac{hc}{\lambda_0 kT} = \frac{6.6 \times 10^{-34}[\text{J s}] \times 3 \times 10^8[\text{m/s}]}{3 \times 10^{-6}[\text{m}] \times 1.38 \times 10^{-23}[\text{J/K}] \times 300[\text{K}]}$$

$$= 15.9\,[\text{unitless}]$$

$$e^{-15.9} = 1.19 \times 10^{-7}$$

$$\lambda_0^{-4} = \left(3 \times 10^{-6}[\text{m}]\right)^{-4} = 1.23 \times 10^{22}\,\text{m}^{-4}$$

$$\overline{\Delta L_p^2} = L_p = 6 \times 10^8[\text{m/s}] \times 2 \times 10^{-6}[\text{m}] \times 1.23 \times 10^{22}[\text{m}^{-4}]$$
$$\times 1.19 \times 10^{-7}$$
$$= 1.24 \times 10^{18}\,\text{s}^{-1}\,\text{m}^{-2}\,\text{sr}^{-1}$$

$$\frac{hc}{\lambda\tau_e} = \frac{6.60 \times 10^{-34}[\text{J s}] \times 3 \times 10^8[\text{m/s}]}{3 \times 10^{-6}[\text{m}] \times 0.19} = 3.47 \times 10^{-19}\,\text{J}$$

$$\frac{2\,\Delta f L_p}{Q\Upsilon} = \frac{2 \times 2.6 \times 10^3[\text{s}^{-1}] \times 1.24 \times 10^{18}[\text{s}^{-1}\,\text{m}^{-2}\,\text{sr}^{-1}]}{0.5 \times 2.4 \times 10^{-3}[\text{cm}^2\,\text{sr}] \times 1 \times 10^{-4}[\text{m}^2/\text{cm}^2]}$$
$$= 5.38 \times 10^{28}\,\text{s}^{-2}\,\text{m}^{-4}$$

$$\left(\frac{2\,\Delta f L_p}{Q\Upsilon}\right)^{1/2} = 2.32 \times 10^{14}\,\text{s}^{-1}\,\text{m}^{-2}$$

$$NES|_{\lambda=3.0\,\mu m} = 3.47 \times 10^{-19}[\text{J}] \times 2.32 \times 10^{14}[\text{s}^{-1}\,\text{m}^{-2}]$$
$$= 8.05 \times 10^{-5}\,\text{W m}^{-2}\,\text{sr}^{-1}$$
$$= 8.05 \times 10^{-9}\,\text{W cm}^{-2}\,\text{sr}^{-1}$$

Note: This problem illustrates that approximately 2 orders of magnitude improvement in NES at 3 μm is obtained by spectral band limiting with an interferometer.   ■

It is common to operate photon detectors in the configuration of Fig. 16.1, where the detector is surrounded by a cold shield, except for the field of view θ, which is exposed to ambient conditions. Typically the theoretical D^* is calculated using Eq. (16.20) and assuming $T = 300$ K. The following is an example of an InSb photovoltaic detector with a cutoff wavelength $\lambda_c = 5.2$ μm when operated at 77 K as shown in Fig. 16.2(b).

Example 3:   Find the theoretical D^* at the cutoff wavelength of 5.2 μm for an InSb photovoltaic detector.

Given: The quantum efficiency is assumed to be unity, and $\theta = 90°$ (full hemisphere of background).

Basic equation:

$$D^*(\lambda) = \frac{\lambda}{2c}\left(\frac{Q}{kTh\pi}\right)^{1/2}\exp\left(\frac{hc}{2kT\lambda_c}\right)$$

$$\left[2\left(\frac{kT}{hc}\right)^2 + \frac{2kT}{hc\lambda_c} + \frac{1}{\lambda_c^2}\right]^{-1/2}(\sin\theta)^{-1}\quad(16.20)$$

Units: cm Hz$^{1/2}$/W

Solution:

$$\lambda = \lambda_c = 5.2 \times 10^{-6}\text{ m}$$

$$2c = 2 \times 3 \times 10^8\text{ m/s}$$

$$\frac{kT}{hc} = \frac{1.38 \times 10^{-23}[\text{J/K}] \times 300[\text{K}]}{6.6 \times 10^{-34}[\text{J s}] \times 3 \times 10^8[\text{m/s}]} = 2.09 \times 10^4\text{ m}^{-1}$$

$$\left(\frac{kT}{hc}\right)^2 = 4.37 \times 10^8\text{ m}^{-2}$$

$$\frac{2kT}{hc\lambda_c} = \frac{2 \times 2.09 \times 10^4[\text{m}^{-1}]}{5.2 \times 10^{-6}[\text{m}]} = 8.04 \times 10^9\text{ m}^{-2}$$

$$\frac{1}{\lambda_c^2} = \frac{1}{5.2 \times 10^{-6}[\text{m}]} = 3.70 \times 10^{10}\text{ m}^{-2}$$

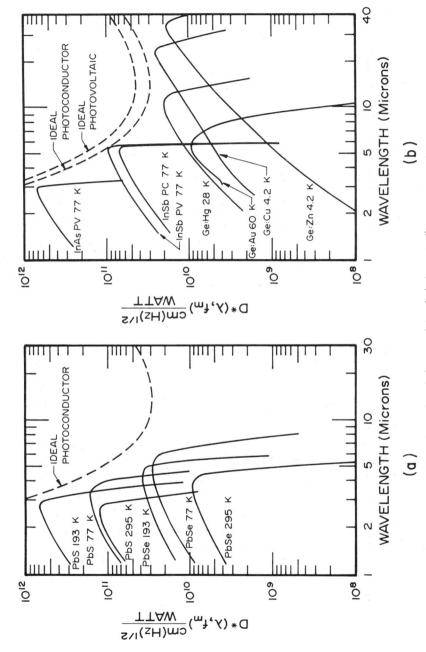

Figure 16.2 Detectivity as a function of wavelength for a variety of photon-counting detectors. Note: Values given are for a 295-K background and π sr field of view. *(Courtesy of Santa Barbara Research Center.)*

245

where the term in square brackets in Eq. (16.20) is

$$[\ ] = 2 \times 4.37 \times 10^8 [\text{m}^{-2}] + 8.04 \times 10^9 [\text{m}^{-2}]$$
$$+ 3.70 \times 10^{10} [\text{m}^{-2}]$$
$$= 4.59 \times 10^{10}\ \text{m}^{-2}$$
$$[\ ]^{-1/2} = 4.67 \times 10^{-6}\ \text{m}$$

$$\frac{Q}{kTh\pi} = \frac{1}{1.38 \times 10^{-23} [\text{J/K}] \times 300 [\text{K}] \times 6.6 \times 10^{-34} [\text{J s}] \times 1.0 \times \pi}$$
$$= 1.16 \times 10^{53}\ \text{J}^{-2}\ \text{s}^{-1}$$

where $Q = 1$, and thus

$$\left(\frac{Q}{kTh\pi}\right)^{1/2} = 3.41 \times 10^{26}\ \text{J}^{-1}\ \text{s}^{-1/2}$$

$$\frac{\lambda}{2c} = \frac{5.2 \times 10^{-6} [\text{m}]}{2 \times 3 \times 10^8 [\text{m/s}]} = 8.67 \times 10^{-15}\ \text{s}$$

$$\frac{hc}{2kT\lambda_c} = \frac{6.6 \times 10^{-34} [\text{J s}] \times 3 \times 10^8 [\text{m/s}]}{2 \times 1.38 \times 10^{-23} [\text{J/K}] \times 300 [\text{K}] \times 5.2 \times 10^{-6} [\text{m}]}$$
$$= 4.6\ [\text{unitless}]$$

$$e^{4.6} = 99.35$$

$$D^*|_{5.2\ \mu\text{m}} = 8.65 \times 10^{-15} [\text{s}] \times 3.41 \times 10^{26} [\text{J}^{-1}\ \text{s}^{-1/2}]$$
$$\times 99.35 \times 4.67 \times 10^{-6} [\text{m}]$$
$$= 1.37 \times 10^9\ \text{m Hz}^{1/2}/\text{W} = 1.37 \times 10^{11}\ \text{cm Hz}^{1/2}/\text{W}$$

Note: The solution to this problem agrees with Fig. 16.2 for InSb (5.2 μm) "ideal photovoltaic" detector. ∎

16.4 TESTING OF PHOTON DETECTORS

Most manufacturers of detectors[4] rate their products in terms of D^*. Several methods are used to measure D^*. It is often advisable to test detectors for compliance with specifications.

One method used to measure $D^*(\lambda)$ is to allow a monochromatic beam of radiation at wavelength λ to fall upon the detector. However, it is difficult to determine the sterance [radiance] of a monochromatic beam. A more common method is to measure the blackbody $D^*(T)$, usually with a 500 K source, and then calculate $D^*(\lambda)$. The T in the parameter $D^*(T)$ could stand for the temperature of the source used or for "total," since it is based upon the *total* blackbody flux.

The blackbody method of measuring $D^*(T)$ requires that two measurements be made: (1) the response $\mathscr{R}(T)$ in V/W to the blackbody flux Φ in W and (2) the noise voltage V_n in V. The detectivity is given by

$$D^*(T) = \mathscr{R}(T)(A_d\,\Delta f)^{1/2}/V_n \qquad [\text{cm Hz}^{1/2}/\text{W}] \qquad (16.24)$$

where $\mathscr{R}(T)$ is given by

$$\mathscr{R}(T) = V_s / \int_0^\infty \Phi_e(\lambda)\, d\lambda \qquad [\text{V/W}] \qquad (16.25)$$

The integral of Eq. (16.25) represents the total incident radiant flux at all wavelengths—including those that are not effective in producing an output signal. Thus, the resulting D^* is not as good as that obtained if only the effective flux had been used. However, the total flux from a blackbody is easily obtained from the Stefan-Boltzmann equation

$$M_e = \sigma T^4 \qquad [\text{W/cm}^2] \qquad (16.26)$$

The value of D^* at the peak (cutoff) wavelength is designated as $D^*(\lambda_c)$ and can be calculated provided the relative spectral response, $\mathscr{R}(\lambda)$, is known. The ratio of $D^*(\lambda_c)$ to $D^*(T)$ is given by the ratio of the responsivities, since the detector area, noise bandwidth, and noise voltage are the same in each case. This ratio equals the ratio of the total flux to the effective flux for a given signal voltage:

$$\frac{D^*(\lambda_c)}{D^*(T)} = \frac{\mathscr{R}(\lambda)}{\mathscr{R}(T)} = \frac{\int_0^\infty \Phi_e(\lambda)\, d\lambda}{\int_0^{\lambda_c} \mathscr{R}(\lambda)\Phi_e(\lambda)\, d\lambda} = a \qquad (16.27)$$

where λ_c is the detector cutoff wavelength and the factor a can be used to calculate $D^*(\lambda)$ where

$$D^*(\lambda_c) = aD^*(T) \qquad (16.28)$$

The value of the ratio of the total flux to the effective flux of Eq. (16.27) can be obtained assuming ideal photon-counting detector response, which is a linear function of wavelength.

$$\mathscr{R}(\lambda) = \lambda/\lambda_c \qquad (16.29)$$

where λ_c is the detector cutoff wavelength. Equation (16.27) can be approximated using numerical methods, combining Eqs. (16.1), (16.26), and (16.29) as follows: The numerator of Eq. (16.27) is given by

$$\int_0^\infty \Phi_e(\lambda)\, d\lambda = \sigma T^4/\pi \qquad (16.30)$$

where Eq. (16.26) is used. The denominator is given by

$$\int_0^{\lambda_c} \mathscr{R}(\lambda)\Phi_e(\lambda)\, d\lambda = \sum_{L=i}^n \frac{\lambda}{\lambda_c} \times \frac{2hc^2}{\lambda^5} \times \frac{\Delta\lambda}{\exp(hc/\lambda kT) - 1} \qquad (16.31)$$

where Planck's equation and Eq. (16.29) are used, and where $\Delta\lambda$ is the (uniform) wavelength increment over n. Thus,

$$a = \frac{\sigma T^4/\pi}{\dfrac{2hc^2\,\Delta\lambda}{\lambda_c} \displaystyle\sum_i^n \lambda_i^{-4}\left[\exp\left(\dfrac{hc}{\lambda_i kT}\right) - 1\right]^{-1}} \qquad (16.32)$$

The solution of Eq. (16.32) is illustrated in Fig. 16.3 for $T = 500$ K.

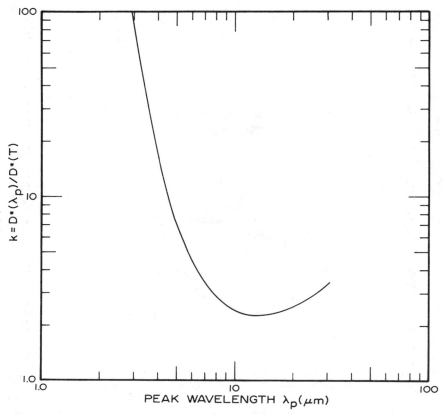

Figure 16.3 Dependence of factor a on detector cutoff wavelength for $T = 500$ K.

EXERCISES

1. List, in order of fundamental considerations, noise mechanisms that limit system noise equivalent power (NEP). What is the most fundamental noise mechanism beyond which no improvement can ever be made? What is the second, and so on?
2. What three methods are available to reduce the background flux incident upon a detector, with given throughput, in order to reduce the photon noise?
3. Cold background operation of photon detectors usually reduces the photon noise to essentially zero values. What other noise mechanism might then limit the system?

For Problems 4 through 8:
Given: An InSb (indium antimonide) detector that is operated photovoltaically at 77 K (liquid nitrogen temperature) with a cold shield as illustrated in Fig. 16.1. The cutoff wavelength is 5.2 μm.

4. Find the theoretical D^* for $\Theta = 90°$ (hemispherical field of view) and background at 300 K. (This corresponds to no cold shield.)

5. Find the theoretical D^* for an $F = 2$ cone cold shield (Fig. 16.1) and background at 300 K.
6. Find the theoretical D^* for an $F = 2$ cone (Fig. 16.1) and background at 300 K that is filtered with a 5.0-μm cold filter ($\Delta\lambda = 0.2$ μm). (Assume an ideal filter.)
7. Find the theoretical D^* for the InSb detector with a 77 K background and $\Theta = 90°$.
8. Find D^*, given that the InSb detector is thermal resistance noise limited. $R_d = 2 \times 10^9$ ohms at 77 K, and the current responsivity is $\mathcal{R}_c = 2.0$ A/W, the detector diameter is 1 mm, and $\Delta f = 1$ Hz.
9. A GeHg (mercury-doped germanium) detector is operated photoconductively at 28 K with its full hemispherical field of view at 28 K. Find the theoretical D^* for $\lambda_c = 12$ μm, given $Q = 0.5$.
10. A PbS (lead sulfide) detector has a cutoff wavelength of 4.0 μm and is operated at 77 K. It is tested with a 500 K blackbody source and is found to have $D^*(T) = 5.6 \times 10^9$ cm Hz$^{1/2}$/W. Find $D^*(\lambda = 4.0$ μm). Check your solution using the graph given in Fig. 16.3. *Note:* Assume linear response, Eq. (16.29), with cutoff wavelength at $\lambda_c = 4.0$ μm. Use Eq. (16.32).
11. Given a beam of monochromatic radiation at 10 μm: the beam power is 1×10^{-12} W. Find the rms power due to fluctuations in the beam. Find the SNR assuming $Q = \Delta f =$ unity.
12. Repeat Exercise 11 for a monochromatic beam at 10 μm where the power is 10^{-16} W.

REFERENCES

1. P. W. Kruse, L. D. McGlauchlin, and R. B. McQuistan, *Elements of Infrared Technology:* Wiley, New York, 1962, p. 361.
2. Ibid., p. 27. Also W. L. Wolf, Ed., *Handbook of Military Infrared Technology*, Office of Naval Research, Dept. of the Navy, Washington, DC, 1965, p. 513.
3. W. L. Wolf, Ed., *Handbook of Military Infrared Technology*, Office of Naval Research, Dept. of the Navy, Washington, DC, 1965, p. 515.
4. A. T. Stair, Jr. et al., "Altitude Profiles of Infrared Radiance of O_3 (9.6 μm) and CO_2 (15 μm), *Geophysical Res. Letters*, **1**, 117 (July 1974).
5. C. L. Wyatt, D. J. Baker, and D. G. Frodsham, "A Direct Coupled Low Noise Preamplifier for Cryogenically Cooled Photoconductive IR Detectors," *Infrared Physics*, **14**, 165–176 (1974).
6. H. Sakai, "Consideration of the Signal-to-Noise Ratio in Fourier Spectroscopy," *Aspen Int. Conf. Fourier Spectry.*, Spec. Rep. No. 114, AFCRL-17-0019, p. 28. Air Force Cambridge Res. Lab., L. G. Hanscom Field, Bedford, MA, 1970.

chapter 17

Multiplier Phototube Detectors

17.1 INTRODUCTION

Photon detectors, unlike thermal detectors, do not depend on temperature; therefore, the thermal capacity has no effect upon their operation. Their response depends upon events such as the electron emission caused by absorption of single quanta of radiation.

17.2 THE PHOTOEMISSIVE EFFECT

Hertz discovered the photoemissive effect in 1887, and Einstein explained it in 1905.[1] Einstein proposed the existence of quanta, each having a discrete amount of energy equal to the product $h\nu$, where h is Planck's constant and ν is the frequency of the radiation. Einstein termed the quanta "photons."

The Einstein equation expresses the phenomenon of photoemission in a concise way as

$$h\nu = E_w + \tfrac{1}{2}mv^2 \quad [\text{J}] \tag{17.1}$$

where h = Planck's constant
   ν = frequency of the radiant energy
   E_w = energy of the surface work function
   m = mass of the electron
   v = velocity of the electron after emission

The electrons are emitted in response to the absorption of a quantum $h\nu$ of energy. The kinetic energy of the emitted electron corresponds to energy in excess of that required to overcome the surface work function. Solving Eq. (17.1) for $v = 0$ corresponds to the frequency of light for which the emitted electrons have zero energy:

$$\nu_0 = E_w/h \quad [\text{Hz}] \qquad (17.2)$$

The corresponding wavelength is given by

$$\lambda_0 = c/\nu_0 \quad [\text{m}] \qquad (17.3)$$

where c is the velocity of light in m/s.

The general laws of photoemission are

1. The kinetic energy of the emitted electron is a function of the frequency or wavelength of the light only; thus, the photoelectron energy is independent of the intensity of the excitation light.

2. The number of electrons emitted is proportional to the intensity or number of photons of the radiant energy.

These laws of photoemission are interpreted as follows: Photons at frequencies less than ν_0 or at wavelengths greater than λ_0 produce no photoelectrons regardless of the excitation light intensity. The ideal responsivity $\mathscr{R}$, with units of amperes per watt, is a linear function of frequency or wavelength.

Emissive surfaces are available that have a variety of surface work functions resulting in the different responsive curves shown in Fig. 17.1.

17.3 PHOTOEMISSION RESPONSIVITY

The photoelectric effect is illustrated with reference to Fig. 17.2. Monochromatic light falling upon the cathode C will liberate photoelectrons, which can be detected as a current if they are attracted to the plate P by means of a potential difference applied between C and P. The galvanometer G serves to measure this photoelectric current. Figure 17.3 is a plot of the photoelectric current in a photodiode like that of Fig. 17.2. The current reaches a limit if the potential is made large enough to collect all the photoelectrons emitted from the cathode. However, the current does not stop even if the potential is reversed. This proves that electrons are emitted with enough excess velocity to reach the plate P in spite of the reverse potential. However, if the reverse potential becomes large enough, it reaches a stopping potential V_0 at which the photoelectric current goes to zero. The product of the potential V_0 and the electronic charge gives the kinetic energy, in electron volts, of the most energetic electrons. The value of the stopping potential is independent of the intensity of the light as shown by curves a and b in which the intensity is related by the factor 2. However, the stopping potential is a function of the frequency of the incident light.

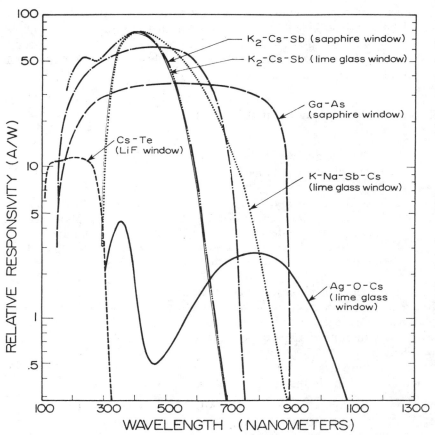

Figure 17.1 Typical spectral response characteristics for multiplier phototube. (From *Practical Photocathode Materials*, courtesy of RCA, New Products Division, Lancaster, PA.)

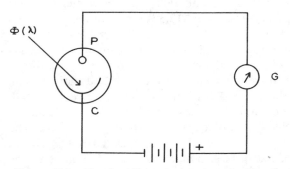

Figure 17.2 Circuit to illustrate the photoelectric effect.

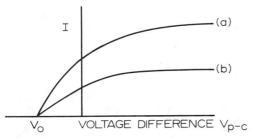

Figure 17.3 Plot of the photoelectric current in a photodiode.

The quantum efficiency Q is defined as the ratio of the number of observable events (photoelectrons) to the number of incident quanta. Ordinarily, the quantum efficiency is ≤ 0.4. The monochromatic power Φ_e incident upon the cathode of a photodetector can be expressed as

$$Nh\nu = \Phi_e \quad [\text{W}] \tag{17.4}$$

where N is the number of incident photons per second. Equation (17.4) can be solved for N to yield

$$N = \Phi_e/h\nu = \Phi_e\lambda/hc \quad [\text{s}^{-1}] \tag{17.5}$$

The number of photoelectrons per second that result from the above quanta is given by

$$n = Q\Phi_e\lambda/hc \quad [\text{s}^{-1}] \tag{17.6}$$

where Q is the quantum efficiency. The cathode current i_c resulting from the photoelectrons is given by

$$i_c = en = eQ\Phi_e\lambda/hc \quad [\text{A}] \tag{17.7}$$

where e is the charge on an electron. The cathode responsivity $\mathscr{R}_c$ is

$$\mathscr{R}_c = i_c/\Phi_e = eQ\lambda/hc \quad [\text{A}/\text{W}] \tag{17.8}$$

Equation (17.8) can be expressed in terms of the atomic constants (see Appendix C) as

$$\mathscr{R}_c = 8.06 \times 10^{-5}Q\lambda \quad [\text{A}/\text{W}] \tag{17.9}$$

where the quantum efficiency Q at wavelength λ is expressed as a decimal and the wavelength λ is expressed in angstroms (10^{-10} m).

17.4 MULTIPLIER PHOTOTUBES

American Standard—Methods of Testing Electron Tubes establishes "multiplier phototube" as the standard term for photoemissive detectors that make use of electron multipliers.[2]

The RCA 4516 is a typical multiplier phototube that consists of an evacuated glass envelope with a Corning No. 0080 lime glass window.[3] The transparent bialkali (K_2-Cs-Sb) photocathode is deposited directly on the inner surface of the 1.5-in.-diameter window. The spectral response is illustrated in Fig. 17.1 and is determined by the transmittance of the window and the relative response of the photocathode. For example, an RCA Type 4522 multiplier phototube uses a fused silica window to obtain improved response in the ultraviolet. Various cathode sizes are available from $\frac{1}{2}$-in. to 5-in. diameter with a variety of spectral responsivities.

The electrons emitted from the cathode in response to incident photons provide a signal current that is amplified in the electron multiplier. Amplification occurs as a result of secondary electron multiplication at one or more electrodes called dynodes. The impact of a primary electron results in the emission of m secondary electrons. The current gain of an n-stage multiplier is given by

$$G = m_1 \times m_2 \times \cdots \times m_n \tag{17.10}$$

where m_1 through m_n are the gains of each stage. The gain per stage is defined as the ratio of current leaving a stage to the current entering that stage. The gain per stage depends upon the secondary emission surface used, the structure of the dynodes, and the interdynode voltage. Thus, the anode current i_A is

$$i_A = i_c G \tag{17.11}$$

where i_c is the cathode current.

Figure 17.4 shows a typical wiring schematic for the type 4516 multiplier phototube. The potential between adjacent stages is equal except for the cathode-to-first-dynode voltage, which is double that of the succeeding interelectrode voltages. This arrangement produces the optimum signal-to-noise ratio (SNR) by reducing secondary emission shot noise. In this case, it is the cathode current shot noise that dominates.

One of the outstanding features of a multiplier phototube is the very fast response time. The limiting frequency to which the multiplier photo-

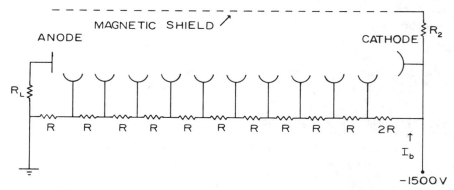

Figure 17.4 Typical wiring schematic for a 10-dynode multiplier phototube.

tube will respond is determined by the transit time spread of the photoelectrons as they traverse their complicated paths, via the dynodes, from cathode to anode. The spread arises from the fact that the path length is not the same for electrons released from different parts of the cathode by the same light signal.

The transit time of an electron leaving the cathode with zero velocity and arriving at the anode with velocity v is given by $t = 2s/v$, where s is the distance between cathode and anode, assuming uniform acceleration. The final velocity v is determined by the anode-to-cathode potential V, according to $v = (2eV/m)^{1/2}$, where e and m are the charge and mass of an electron, respectively. The spread in transit time is therefore

$$\Delta t = 3.4 \times 10^{-8} \Delta s / V^{1/2} \quad [\text{s}] \qquad (17.12)$$

where the cathode-to-anode separation s is in cm.

In a typical phototube, the path length from the cathode to the anode may vary by as much as 1 cm. If the supply voltage is 300 V, the transit time spread is about 2 ns. The frequency response of such a phototube extends from dc to about 500 MHz.

In a multiplier phototube, a much greater transit time spread occurs because electrons released from the various parts of each dynode will have differences in their path lengths to the next dynode. In a 10-dynode multiplier, the upper frequency limit drops to about 100 MHz.

The lapse time between the signal in and the signal out in a vacuum tube device is the time required for an electron to travel from one electrode to another. The lapse time is only a fixed delay and has no appreciable effect on the frequency response of the device.

Practical high-performance multiplier phototubes have recently become available for use in high-data-rate laser communication systems.[4] High-speed laser systems under development require 1- to 3-GHz bandwidth detectors. Computer-optimized dynode structures using wide-band microwave collector designs have resulted in several types. These designs employ fewer dynodes with higher interelectrode voltages to achieve the required gain and speed.

A frequency response of 10 GHz has been achieved in designs that use an rf field in a continuous multiplier. The rf field "bunches" the electrons to eliminate those traveling over different paths or those emitted out of synchronization with the radio frequency. These detectors have found applications, other than for high-data-rate communications, that capitalize upon the sensitive and fast detection.

Multiplier phototubes were initially used to measure the visual effect and are still tested and specified in terms of luminous flux. The lumen is defined[5] in terms of a 2040 K blackbody and a filter that has a "standard" eye spectral transmittance V_λ. The coefficient that relates lumens to watts and recommended by the National Bureau of Standards is 680 lm/W.

17.4.1   Noise in Multiplier Phototubes

The figure of merit D^* is not appropriate for multiplier phototubes. Such a rating is based upon an area dependence that is not justified in this case. The noise in a multiplier phototube output signal may be given a broad classification as originating from either photon-excited electron emission or nonphoton origin.

Following IEEE standards,[2] the noise output current fluctuations for any unmodulated input flux (either background or signal) are called "noise in signal." Noise in signal is present only when flux, within the spectral responsive wavelength region, is incident upon the photocathode.

Thus, the observed noise increases whenever flux is incident upon the cathode of a multiplier phototube. This is the result of the statistical fluctuations of photoemission, which are modified by gain variations in the multiplier. In many applications, such as the detection of a point source, there is a background flux contributing to the noise, in addition to the signal. This additional noise component might be termed "background" noise, and it often predominates in many applications. In any case, it is important to include all noise forms in determining the SNR.

There is always a residual output current in the absence of any flux incident upon the cathode; this is termed "dark noise." Dark noise is caused by several mechanisms: thermionic emission from the cathode, which, like noise in signal, has a distribution independent of frequency, and unpredictable noise sources such as leakage current, dynode emission, gas discharge, ion drift, and radioactive excitation. The dark noise can be measured by simply covering the cathode with an optically opaque mask, since phototubes normally do not respond to thermal radiation.

Background noise, noise in signal, and photocathode thermionic noise can be predicted by means of the cathode dc dark current i_c, using the shot-noise relationship

$$\overline{i_n^2}(\text{anode}) = 2eG^2ki_c\,\Delta f = 2eki_AG\,\Delta f \quad [\text{A}^2] \quad (17.13)$$

where i_n is the anode rms noise current, e is the electronic charge, G is the multiplier gain, k is the multiplier noise factor, i_A is the anode dc dark current, and Δf is the noise bandwidth. The total noise is the sum of the squares for the various components considered.

For discrete-stage electron multipliers, which obey Poisson statistics for the secondary emission process, the noise factor k can be approximated as

$$k = m/(m - 1) \quad (17.14)$$

where m is the average gain per stage.

The noise in multiplier phototube photoemission current, both thermionic and photoemissive, is known to be independent of frequency from dc to the upper frequency limit set by the transit time. On the other hand, noncathode noise sources do not exhibit a uniform frequency distribution and cannot be predicted.

Leakage currents occur between electrodes in the tube socket, base, stem, and internal parts. For example, a leakage resistance of 10^{12} ohms at 1 kV potential yields 10^{-9} A current—a relatively large signal. The noise resulting from leakage currents would be negligible if it were purely ohmic; however, it is characterized by sudden discharges and is very erratic. Leakage noise is often the limiting noise form except when pulse-counting techniques are used.

Thermionic noise in multiplier phototubes is more serious in cathodes that possess a lower surface work function and are therefore more responsive into the red part of the spectrum. This form of noise can be reduced by cooling the photomultiplier. The dark noise for type S–1 tubes, for example, falls about an order of magnitude per 10°C of cooling[6]; however, it varies with different tube designs. There are some hazards associated with cooling multiplier phototubes. Structural failure may occur in the graded seals between the glass base and the window. Also, it has been shown that cooling reduces the thermionic noise down to approximately 0°C.

Ion drift can occur in the glass envelope when a large electric field exists between the electrodes and an external shield. This is especially serious when such a potential exists in the region of the cathode. Ion drift through the window can actually destroy the photemissive surface. Generally, this problem can be avoided by insulting the tube from the shield with a layer of mylar and connecting the magnetic shield to the cathode potential through a large isolation resistor as Fig. 17.4 shows.

The superior performance obtained using pulse-counting techniques depends on pulse-height discrimination. Pulses originating from cathode emission experience greater multiplication than other forms of dark current noise.

17.4.2 Responsivity of a Multiplier Phototube

The anode responsivity $\mathscr{R}_A$ is given in terms of the cathode responsivity $\mathscr{R}_c$ by Eq. (17.9), the gain of Eq. (17.10), and Eq. (17.11) as

$$\mathscr{R}_A = \mathscr{R}_c G \quad [\text{A/W}] \tag{17.15}$$

where G is the multiplier gain.

The multiplier phototube can be used in any of three modes: (1) dc-coupled, (2) ac-coupled with a light chopper, or (3) pulse counting.

17.4.3 Direct-Coupled Noise Equivalent Power

The direct-coupled noise equivalent power (NEP) is defined as the incident power that results in a change in the output equal to the multiplier phototube dark current. This is given by

$$\text{NEP}_{\text{dc}} = i_A(\text{dc})/\mathcal{R}_A \quad [\text{W}] \qquad (17.16)$$

where i_A is the anode dc dark current. Combining Eqs. (17.9), (17.11), (17.15), and (17.16),

$$\text{NEP}_{\text{dc}} = i_c/\left[8.06 \times 10^{-5} Q\lambda(A)\right] \quad [\text{W}] \qquad (17.17)$$

where i_c is the cathode dark current.

17.4.4   Equivalent Noise Input

Following IEEE standards,[2] the noise equivalent flux (NEF) for a multiplier phototube operated with a light chopper is termed *equivalent noise input* (ENI) rather than NEP.

A light chopper provides an alternating electric signal output from the multiplier phototube. The narrow-band ac noise may depend upon the electrical bandwidth. In this case, the rms noise voltage may be significantly lower than the dc dark current, and light chopping may therefore improve the ability to detect faint signals.

The ac noise is proportional to the square root of bandwidth only when the dark noise is uniform with frequency, or "white." Generally this is true for multiplier tubes only where they are thermal-noise-limited. If the dominant noise is leakage, or otherwise colored, chopping may not improve performance. In this case, pulse-counting techniques are superior.

When the dark noise is proportional to the square root of bandwidth, the equivalent noise input is given by

$$\text{ENI} = \left(2 e i_c \Delta f\right)^{1/2}/\mathcal{R}_c \quad [\text{W}] \qquad (17.18)$$

where e is the charge on an electron (1.6×10^{-19} C), i_c is the cathode dc dark current, Δf is the electrical noise bandwidth, and $\mathcal{R}_c$ is the cathode responsivity given by Eq. (17.9).

Example 1:   Find the NEP(dc) and ENI for a type 4516 bialkali multiplier phototube detector.

Given: Quantum efficiency is 0.24 at 4000 Å. Anode dark current is 2×10^{-10} A at multiplier gain of 4×10^5.

Basic equations:

$$\mathcal{R}_c = 8.06 \times 10^{-5} Q\lambda(A) \quad [\text{A/W}] \qquad (17.9)$$

$$\text{NEP}_{\text{dc}} = i_A(\text{dc})/\mathcal{R}_A \quad [\text{W}] \qquad (17.16)$$

$$\text{ENI} = \left(2 e i_c \Delta f\right)^{1/2}/\mathcal{R}_c \quad [\text{W}] \qquad (17.18)$$

Assumptions: The noise is white, and $k = 1$.

Solution: Cathode responsivity is given by

$$\mathscr{R}_c = 8.06 \times 10^{-5} Q\lambda(A) = 8.06 \times 10^{-5} \times 0.24 \times 4000$$
$$= 0.0775 \text{ A/W}$$

The cathode dark current is given by

$$\dot{i}_c = i_A/G = 2 \times 10^{-10}/4 \times 10^5 = 5 \times 1^{-16} \text{ A}$$

The dc NEP is

$$\text{NEP} = i_c/\mathscr{R}_c = 5 \times 10^{-16}/0.0775 = 6.45 \times 10^{-15} \text{ W}$$

The ac equivalent noise input is

$$\text{ENI} = \frac{(2ei_c)^{1/2}}{\mathscr{R}_c} = \frac{(2 \times 1.6 \times 10^{-19} \times 5 \times 10^{-16})^{1/2}}{0.0775}$$
$$= 1.63 \times 10^{-16} \text{ W/Hz}^{1/2}$$

System NEF is degraded by the chopping factor (typically 3) and the coherent rectification factor $\sqrt{2}$; thus, the system ENI is

$$\text{ENI(sys)} = 1.63 \times 10^{16} [\text{W Hz}^{1/2}] \times 3 \times \sqrt{2} = 6.92 \times 10^{-16} \text{ W/Hz}$$

which indicates an improvement of about one order of magnitude.   ∎

Table 17.1 provides calculations for a number of tube types based upon the method of Example 1.

17.4.5 Pulse Counting

Pulse-counting techniques capitalize upon the fact that photoemission from the cathode experiences a greater gain, thus resulting in a greater pulse amplitude in the output, than pulses that have their origin in photoemission from a dynode or interelectrode leakage. Using pulse-height discrimination circuits eliminates all pulses not originating from the cathode. In practical application, those pulses that exhibit an amplitude above some threshold value are assumed to originate in the photocathode and are counted. Typically, the counter is gated so that the output is the average count over a period τ.

Under ideal conditions, where the multiplier phototube is limited by photon noise in the signal, there are no advantages to pulse-counting techniques. In addition, it is not possible to distinguish between thermionically emitted electrons and true photoelectrons. In either case, the mean-square deviation in the count is exactly the same as for the current in an analog system for the same observation time. In Chap. 7, it was pointed out that for independent events such as the emission of a photoelectron, the statistical distribution is governed by the Poisson function. The Schottky equation [Eq. (7.29)] gives the mean-square fluctuations in the current.

Table 17.1 NEP AND ENI FOR 1-Hz BANDWIDTH FOR TYPICAL MULTIPLIER PHOTOTUBES

RCA type	Photocathode type	Window material	Wavelength, Å	Q, %	$\mathscr{R}_c$, A/W	i_c, A	NEP, W	ENI, W/Hz$^{1/2}$
C70128	CsTe	Lithium fluoride	200	3.0E-3	5.0E-4	1.0E-18	2.0E-15	1.1E-15
4522	K_2-Cs-Sb	Fused silica	400	22.0	7.10E-2	2.86E-14	4.08E-13	1.37E-14
4516	K_2-Cs-Sb	Lime glass	4000	24.5	7.90E-2	5.00E-16	6.33E-15	1.64E-16
8644 (S-20)	K-Na-Cs-Sb	Lime glass	4000 7000	19.0 2.5	6.13E-2 1.40E-2	7.50E-15	1.22E-13 5.35E-13	8.03E-16 3.50E-15
7102 (S-1)	Ag-O-Cs	Lime glass	7500 11000	0.42 0.20	2.54E-3 1.77E-3	4.48E-11	1.76E-8 2.53E-8	1.49E-12 2.14E-12

Source: Courtesy of RCA, New Products Division, Lancaster, PA.

For a pulse-counting system, the number of counts n_τ obtained during sample time τ varies from time to time. However, the average number of counts $\overline{n}_\tau$ for a sample time is independent of time (assuming the photon source is stationary in time), and the steady sample rate $\overline{n}_0$ is

$$\overline{n}_0 = \overline{n}_\tau/\tau \quad [\text{s}^{-1}] \tag{17.19}$$

The mean-square deviation in the count rate for a sample time τ is

$$\overline{\Delta n_\tau^2} = \overline{n}_0 = \overline{n}_\tau/\tau \tag{17.20}$$

which is large for small sample times. The rms count rate deviation is given by

$$\Delta n(\text{rms}) = (\overline{n}_0)^{1/2} \tag{17.21}$$

For example, the rms deviation in the count rate for an average of 100 counts/s is ± 10 counts/s or $\pm 10\%$. On the other hand, a count rate of 10^4 counts/s has an rms deviation of only $\pm 1.0\%$. It follows that the SNR is

$$\text{SNR} = \overline{n}_0\tau/(\overline{n}_0\tau)^{1/2} = (\overline{n}_0\tau)^{1/2} \tag{17.22}$$

The sample time τ corresponds to the sample frequency f_s, which must be established in accordance with the sampling theorem as follows: A modulated signal can be completely reproduced provided that the sample frequency is

$$f_s = 2f_m \tag{17.23}$$

where f_m is the maximum frequency for which the signal contains sufficient energy to be significant. It therefore follows that the information bandwidth Δf is

$$\Delta f = f_m = \frac{f_s}{2} = \frac{1}{2\tau} \tag{17.24}$$

The cathode-to-first-dynode voltage is made relatively large so that photoelectrons and/or thermionic electrons experience greater gain than electrons emitted from any dynode. Pulse-counting techniques provide an advantage only when photocathode thermionic electrons have been eliminated by design or by cooling. The ultimate limit to the pulse-counting technique is set by electrons emitted from the cathode that result from cosmic events—gamma rays, secondary beta rays, etc.

EXERCISES

1. Verify Eq. (17.9) using the values of the atomic constants given in the appendix and unit analysis.
2. The dark current of a multiplier phototube is measured across a load resistor R_L. The multiplier has a gain of 10^6. Find the value of R_L that yields thermal noise voltage equal to the shot noise for an anode dark current of 10^{-9} A.

Hint: Assume the dynamic resistance of the multiplier phototube can be neglected. Also assume $k = 1$ and $T = 300$ K.

3. A photometer is required to monitor both amplitude and rate of change of the N_2^+ emission of the active thermosphere. The maximum sterance [radiance] for a Class III aurora is 1×10^{-8} W cm^{-2} sr^{-1}. The aurora contains modulation frequencies up to 100 Hz. A design goal of 10^5 dynamic range is desired with $1°$ full-angle field of view. Given: Photocathode size $\frac{1}{2}$-in diameter, filter 3940 Å, bandwidth 30 Å, optical efficiency 0.2, $Q = 0.24$, $f/2$ optics, tube type 4516 (Table 17.1). Find the NES, dynamic range, and the collector diameter and focal length. *Hint:* Dynamic range $= 1 \times 10^{-8}/$NES.

4. A figure of merit for an amplifier is its equivalent input noise resistance (based upon thermal noise $\sqrt{4kTR}$). Consider the shot noise–limited electron multiplier as an amplifier—find its equivalent input noise resistance. Given: Cathode current 1×10^{-15} A, gain 1×10^6, load resistor 1×10^7 ohms, $T = 300$ K. Compare this with a high-quality operational amplifier for which the input noise voltage is 1×10^{-8} V/Hz$^{1/2}$.

REFERENCES

1. D. Halliday and R. Resnik, *Physics*, Wiley, New York, 1963, p. 1087.
2. *American Standard—Methods of Testing Electron Tubes*, "IRE Standards on Electron Tubes—Methods of Testing—1962" (62 IRE 7 S 1) (IEEE 158) (ASA C60.15-1963), pp. 78–89.
3. Made by Corning Glass Works, Corning, NY, 15830.
4. R. S. Ench and W. G. Abraham, "Review of High Speed Communications Photomultiplier Detectors," *Proc. Soc. Photo-Opt. Instrum. Eng.*, **150**, 40–48 (1978).
5. National Bureau of Standards, U.S. Dept. of Commerce, *Opt. Radiat. News*, **7**, 1–2 (1975).
6. *Threshold Sensitivity and Noise Ratings of Multiplier Phototubes*, Application Note E2, ITT Industrial Laboratories, Ft. Wayne, IN, December 1964.
7. A. T. Young, "Undesirable Effects of Cooling Photo-multipliers," *Rev. Sci. Instrum.*, **38**, 1336 (1967).

chapter *18*

Low-Noise Preamplifiers

18.1 INTRODUCTION

The detector of optical radiant energy is a transducer that produces an electric signal response to incident energy rate or incident photon rate. This chapter deals with circuits and devices utilized to convert the detector signal into a useful voltage level.

The circuit designer is concerned primarily with the volt-ampere (V-I) characteristics of the detector. An optimum design is one that couples the faint electric signal into the amplifier circuit without degrading the detector performance.

18.1.1 Photoconductive Detectors

A photoconductive detector is one for which the resistance is inversely proportional to the photon rate. Thus, a method must be devised to measure the resistance change. This can be accomplished by using a bridge circuit as shown in Fig. 18.1. The change in the detector resistance R_d that occurs when incident photons are absorbed causes an imbalance in the bridge, providing an indication on the galvanometer G. The variable resistor in the right arm of the bridge is adjusted to null the dark signal.

Normally detectors are used with light choppers to avoid $1/f$ noise that is present in semiconductors. In this case, the right arm of the bridge and the galvanometer can be replaced with a capacitor and ac amplifier.

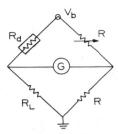

Figure 18.1 Bridge circuit for measuring resistance change of a photoconductive detector.

The capacitor effectively blocks the dark signal and couples only the changing voltage at the junction of R_d and R_L into the amplifier. This is illustrated with the circuit of Fig. 18.2. The bias current is given by

$$I_b = V_b/(R_d + R_L) \quad [A] \qquad (18.1)$$

where R_L is termed the "load resistor" and R_d is the "dark" resistance of the detector. The resulting voltage at point P is obtained for the circuit as

$$V_p = \frac{V_b R_L}{R_L + R_d} \quad [V] \qquad (18.2)$$

Only the change in V_p, for a change in R_d, is coupled into the amplifier. This can be obtained by assuming that the illuminated detector resistance is $R_d - \Delta R$, thus

$$\Delta V_p = \frac{V_b R_L}{(R_d + R_L)^2} \Delta R = \frac{I_b R_L}{R_d + R_L} \Delta R \quad [V] \qquad (18.3)$$

where I_b is the bias current. The output ΔV_p is a linear function of ΔR, provided that $\Delta R \ll R_d$. This equation is often written in the form

$$\Delta V_p = F I_b \Delta R \quad [V] \qquad (18.4)$$

where $F = R_L/(R_d + R_L)$ is called the "bridge factor."[1] For constant bias voltage, the condition $R_d = R_L$ yields the maximum output.

Another circuit for photoconductive detectors is one that terminates the detector in a short circuit. This is possible using an operational amplifier as shown in Fig. 18.3. Feedback at the inverting input provides a

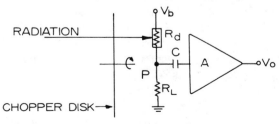

Figure 18.2 Circuit for detector bias and ac amplifier.

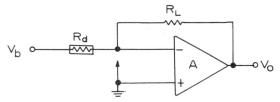

Figure 18.3 Operational amplifier circuit illustrating the virtual ground.

"virtual ground."[2] The output, for an ideal amplifier, is

$$V_o = \frac{-V_b}{R_d} R_L = -I_b R_L \quad [\mathrm{V}] \tag{18.5}$$

Again, only the change in output for chopped radiation is of interest. The change in the output for a change in detector resistance is given by

$$\Delta V_o = -I_b \frac{R_L}{R_d} \Delta R_d \quad [\mathrm{V}] \tag{18.6}$$

The output ΔV_o, is a linear function of ΔR_d as with the bridge circuit, provided $\Delta R_d \ll R_d$.

Photoconductive detectors operated under reduced temperature conditions in the infrared exhibit very high resistance values, particularly when operated under reduced background (enhanced) conditions. Thus, the detector resistance may be several orders of magnitude greater than the load resistor.[3]

18.1.2 Photovoltaic Detectors

Photovoltaic detectors are made from grown or diffuse junctions and can be operated either in the unbiased photovoltaic mode or in the reverse photoconductive bias mode. The highest D^* values are obtained in the unbiased photovoltaic mode, while maximum response rates are obtained for biased detectors.

The V-I curve for a typical InSb photovoltaic detector is given in Fig. 18.4. These data are obtained for the detector cooled to 78 K and shielded at the same temperature from any ambient background. Figure 18.4 is interpreted as follows: The slope, dv_i/di, represents the junction resistance R and has a maximum value at $V = 0$. The detector capacitance is proportional to area and is given approximately by $C = 5 \times 10^{-8}$ F/cm². A figure of merit for photovoltaic detectors is the RA product (where A is the detector area). Limiting values of RA are achieved for given detector types; thus, larger detectors have smaller resistance. Large detector resistance values yield higher D^* values as shown below. For this reason, short-circuit operation ($V = 0$) yields superior performance.

Two $V = 0$ modes are possible: (1) A transformer primary winding provides a low-resistance termination for the detector. The secondary

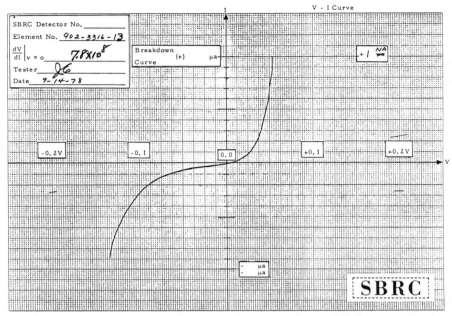

Figure 18.4 The *V-I* curve for a typical InSb photovoltaic detector. (Courtesy of Santa Barbara Research Center.)

winding yields the output voltage. (2) The operational amplifier virtual ground of Fig. 18.3 provides a very low impedance termination for the detector. The output voltage of the current-to-voltage converter is given by

$$\Delta V_o = \Delta i_s R_L \qquad [\text{V}] \qquad (18.7)$$

where Δi_s is detector signal current.

Typical InSb photovoltaic detectors have current responsivities of 2.0 A/W. Silicon diffused junction detectors (0.3 to 1.1 μm) have typical current responsivities of 0.5 A/W.

18.1.3 Photoemissive Detectors

Chapter 17 covered multiplier phototubes in detail. The *V-I* characteristics of a typical multiplier phototube detector[4] are illustrated in Fig. 18.5. The data of Fig. 18.5 are interpreted as follows: The anode current is relatively independent of anode-to-last-dynode voltage provided the voltage is sufficiently high. In the flat region of the curves, the anode resistance $R_d = dv/di > 200$ megohms. Thus, the multiplier phototube behaves as a high-impedance current source.

The basic circuit for a multiplier phototube is given in Fig. 17.4. The ac-equivalent circuit for a multiplier phototube is given in Fig. 18.6, where R_d is the anode-to-last-dynode dynamic resistance (dv/di) and the corre-

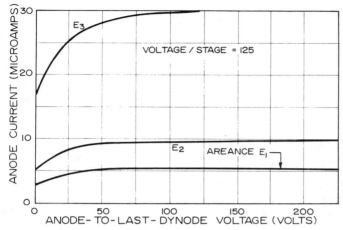

Figure 18.5 Typical *V-I* curves for a multiplier phototube detector.

sponding voltage is ΔV_A. A node equation for point P gives

$$\Delta i_s = \frac{-\Delta V_A}{R_d} + \Delta i_A \quad [\text{A}] \tag{18.8}$$

where Δi_s is the signal current given by the product of the current responsivity $\mathcal{R}_c$ in amperes per watt and the incident flux Φ_e in watts:

$$\Delta i_s = \mathcal{R}_c \Phi_e \tag{18.9}$$

and where ΔV_A represents a decrease in the interelectrode voltage due to the flow of anode current in R_L (see Fig. 17.4) and where Δi_A is the anode current.

Linear operation of the multiplier tube[4] requires that Δi_d be negligible compared to Δi_A, or that

$$\Delta i_s = \Delta i_A \tag{18.10}$$

which occurs when $R_L \ll R_d$. This condition is achieved by using the operational amplifier of Fig. 18.3.

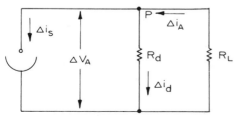

Figure 18.6 Equivalent circuit for a multiplier phototube where ΔV_A and R_d are the last-dynode-to-anode voltage and dynamic resistance (*dv / di*), respectively.

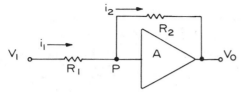

Figure 18.7 Circuit to illustrate the transfer function of the transverse impedance preamplifier.

18.2   TRANSIMPEDANCE PREAMPLIFIER

The operational amplifier of Fig. 18.3 is referred to as a "transimpedance preamplifier."[3] The following analysis of the transfer function and the virtual ground of the feedback amplifier is based upon Fig. 18.7.

The transfer function is defined as

$$T_r = \frac{V_o}{V_1} \tag{18.11}$$

and is obtained by writing a node equation for the point P.

$$i_1 = i_2 \tag{18.12}$$

which is true only if the current flow into the amplifier is negligibly small. An amplifier with an extremely low input current (or high input impedance) is referred to as an "electrometer" amplifier.[5-8] The values of i_1 and i_2 are obtained from the figure:

$$\frac{V_1 + V_o/A}{R_1} = \frac{-V_o/A - V_o}{R_2} \tag{18.13}$$

where the amplifier gain is given by

$$A = -V_o/V_p \tag{18.14}$$

for an inverting operational amplifier. The characteristics of an electrometer can be obtained using field-effect devices at the amplifier input. High-quality junction FETs have input currents as low as 10^{-13} A when operated at reduced temperatures.

Solving Eq. (18.13) for the transfer function yields

$$T_r = \frac{V_o}{V_1} = \frac{-R_2}{R_1}\left(\frac{1}{1 + (1/A)(1 + R_2/R_1)}\right) \tag{18.15}$$

which in the limit as A approaches infinity becomes

$$T_r = \frac{-R_2}{R_1} \tag{18.16}$$

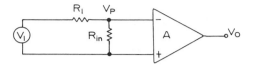

Figure 18.8 Equivalent circuit of the operational amplifier to illustrate the virtual ground.

18.3 VIRTUAL GROUND

The virtual ground at the inverting input of an operational amplifier is obtained using Eqs. (18.12) and (18.14) to get

$$i_1 = i_2 = \frac{V_p - V_o}{R_2} = \frac{V_p(1 + A)}{R_2} \tag{18.17}$$

and the input impedance R_{in} is

$$R_{in} = \frac{V_p}{i_1} = \frac{R_2}{1 + A} \tag{18.18}$$

as illustrated in Fig. 18.8, which approaches zero in the limit as A approaches infinity.

Typical operational amplifiers have dc gains as high as 10^5 resulting in an effective short-circuit termination for R_1.

18.4 FREQUENCY RESPONSE

The above analysis of the transfer function is valid only at low frequencies because the effect of circuit capacitances has been neglected. The purpose of this section is to consider the effect of circuit capacitances upon the transfer function and to introduce frequency-compensation techniques.[3]

The circuit of the operational amplifier and detector-bias network with the associated shunt and input capacitances are shown in Fig. 18.9.

The bias supply is a zero-impedance dc voltage; thus, the circuit of Fig. 18.9 can be simplified by lumping the detector and input capacitances together in C_T as shown in Fig. 18.10, where the detector is represented as a current source.

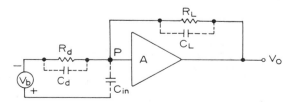

Figure 18.9 The circuit of the operational amplifier and detector bias network.

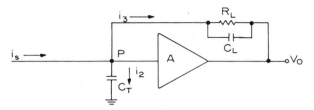

Figure 18.10   Equivalent ac circuit for the operational amplifier.

The transfer function for the circuit of Fig. 18.10 is defined as

$$T_r = V_o/i_s \qquad (18.19)$$

which has the units of ohms, and gives rise to the term "transimpedance" attributed to this current-to-voltage configuration.

The effect of the capacitances at higher frequencies is determined by solving for the transfer function in the s domain, letting i_s become a step function, and determining the system response using the appropriate Laplace transform.[9] We proceed as before to write a node equation for point P:

$$i_s = i_2 + i_3 = V_p s C_T + \frac{(V_p - V_o)(R_L C_L s + 1)}{R_L} \qquad (18.20)$$

and using $V_p = -V_o/A$ to get

$$i_s = \frac{-V_o}{R_L A}[R_L C_T s + R_L C_L s + 1 + (R_L C_L s + 1)A] \qquad (18.21)$$

It is necessary to factor out s to obtain a standard Laplace form:

$$i_s = \frac{-V_o}{R_L A}\{s[R_L C_T + R_L C_L(A + 1)] + (A + 1)\} \qquad (18.22)$$

or

$$i_s = \frac{-V_o[R_L C_T + R_L C_L(A + 1)]}{R_L A}\left(s + \frac{A + 1}{R_L C_T + R_L C_L(A + 1)}\right) \qquad (18.23)$$

Letting i_s become a step function, the output is

$$V_o = \frac{A R_L}{R_L C_T + R_L C_L(A + 1)}\left(\frac{1}{s}\right)\left(s + \frac{A + 1}{R_L C_T + R_L C_L(A + 1)}\right)^{-1} \qquad (18.24)$$

We recognize Eq. (18.24) as a standard transform:

$$\beta\frac{1}{s}\left(\frac{1}{s + \alpha}\right) \rightarrow \beta\frac{1 - \exp(-\alpha t)}{\alpha} \qquad (18.25)$$

where

$$\beta = \frac{AR_L}{R_LC_T + R_LC_L(A + 1)} \qquad (18.26)$$

and

$$\alpha = \frac{A + 1}{R_LC_T + R_LC_L(A + 1)} \qquad (18.27)$$

The time constant $T_c = 1/\alpha$ [Eq. (18.25)] is given by

$$T_c = \frac{R_LC_T + R_LC_L(A + 1)}{A + 1} = R_L\left(\frac{C_T}{A + 1} + C_L\right) \qquad (18.28)$$

Equation (18.28) is interpreted as follows: The input time constant R_LC_T is reduced by the gain, but the load resistor time constant R_LC_L is not.

The open-loop gain A of an operational amplifier is a function of frequency. Typically the gain is 10^5 from dc to 10 Hz and then rolls off at 6 dB/octave. The gain can be expressed as

$$A = \frac{k}{f + a} \qquad (18.29)$$

where a is the frequency of the first pole and k is the gain-bandwidth product given by the product of the dc gain and a; i.e., typically $k = a \times A_{dc} = 10^6$.

The following is an example calculation for an InSb detector based upon the data of Fig. 18.4.

Example 1: Calculate the time constant and frequency response for a 1-mm-diameter InSb detector and an 1×10^9-ohm Eltec load resistor (which has 0.2-pF shunt capacitance) in the circuit configuration of Fig. 18.2. *Note:* No feedback.

Given: The amplifier of Fig. 18.2 is a conventional high-impedance ac amplifier (without feedback) using FET inputs to provide high impedance.

Basic equations:

$$T_c = R_{eq}C_T \quad [\text{s}] \qquad (\text{Time constant})$$
$$C = 5 \times 10^{-8} \text{ F/cm}^2 \qquad (\text{For InSb detectors})$$
$$f_2 = \frac{1}{2\pi T_c} \quad [\text{Hz}] \qquad (7.3)$$

Assumptions: The amplifier is an ideal electrometer.

Solution: The detector of Fig. 18.4 has a $V = 0$ resistance of 7.8×10^8 ohms. The equivalent circuit resistance is the parallel combination of R_d

and R_L, which for $R_l = 10^9$ is

$$R_{eq} = 7.8 \times 10^8 \times 1 \times 10^9/(7.8 \times 10^8 + 1 \times 10^9) = 4.38 \times 10^8$$

The total input capacitance consists of the sum of the detector, stray, and input capacitances. The detector capacitance is

$$C_d = 5 \times 10^{-8}[\text{F/cm}^2] \times 7.85 \times 10^{-3}[\text{cm}^2] = 393 \text{ pF}.$$

The stray capacitance is typically 2 pF, and the FET input capacitance is 6 pF; thus,

$$C_T = 393 + 2 + 6 = 401 \text{ pF}$$

The time constant is

$$T_c = 4.38 \times 10^8 \times 401 \times 10^{-12} = 0.176 \text{ s}$$

and the frequency response is

$$f = 1/(2\pi \times 0.176) = 0.9 \text{ Hz} \qquad \blacksquare$$

The above example illustrates the relatively poor response of the high-impedance detector circuits utilizing conventional ac amplifier techniques. The next example illustrates the frequency-compensating effect of feedback at the input of an operational amplifier electrometer.

Example 2:   Calculate the time constant and frequency response for the InSb detector of Example 1, utilizing the operational amplifier configuration of Fig. 18.10.

Given: A Signetics FET input operational preamplifier type 536 is used. The 536 has the following pertinent typical characteristics: Input bias current is 30 pA, input capacitance is 6 pF, and the dc gain is 10^5. The detector capacitance is 5×10^{-8} F/cm². The feedback resistor $R_L = 1 \times 10^9$ ohms, and $C_L = 0.2$ pF.

Basic equations:

$$T_c = R_L\left(\frac{C_T}{A + 1} + C_L\right) \qquad (18.28)$$
$$A = k/(f + a) \qquad (18.29)$$
$$f_2 = 1/2\pi T_c \qquad (7.3)$$

Assumptions: The amplifier is an ideal electrometer, and at high frequency the gain can be approximated as k/f and $A \gg 1$.

Solution: The time constant is

$$T_c = R_L C_T f/k + R_L C_L$$

and the frequency response is

$$f = 1/2\pi T_c$$

Eliminating the time constant gives

$$R_L C_T f/k + R_L C_L = 1/2\pi f$$

which is rewritten in quadratic form as

$$f^2 + (kC_L/C_T)f - k/2\pi R_L C_T = 0$$

This equation can be solved for f utilizing the quadratic formula

$$f = -\frac{b}{2a} \pm \left[\left(\frac{b}{2a}\right)^2 - \frac{c}{a}\right]^{1/2}$$

as follows:

$$\frac{b}{2a} = kC_L/2C_T = 10^6 \times 0.2 \times 10^{-12}/(2 \times 400 \times 10^{-12}) = 250$$

$$(b/2a)^2 = 250^2 = 6.25 \times 10^4$$

$$\frac{c}{a} = \frac{k}{2\pi R_L C_T} = \frac{1 \times 10^6}{2\pi \times 10^9 \times 400 \times 10^{-12}} = 3.98 \times 10^5$$

$$\left[\left(\frac{b}{2a}\right)^2 - \frac{c}{a}\right]^{1/2} = (6.25 \times 10^4 + 3.98 \times 10^5)^{1/2} = 678.5$$

Thus,

$$f = -250 + 678.2 = 428.5 \text{ Hz}$$

where the positive solution is used.

Assuming $C_L = 0$ the solution is

$$f = \left(\frac{k}{2\pi R_L C_T}\right)^{1/2} = \left(\frac{10^6}{2\pi \times 10^9 \times 400 \times 10^{-12}}\right)^{1/2} = 631 \text{ Hz}$$

which illustrates the limiting effect of the load resistor time constant that is not reduced by the gain in this configuration. ∎

Example 3 illustrates the design considerations for a multiplier phototube application.

Example 3: Determine the load resistor, time constant, frequency response, and system responsivity in volts/watt for a Type 4516 multiplier phototube operated in the dc mode. The 5-ft-long coaxial cable (capacitance 150 pF) is used to connect the detector to the type 536 operational amplifier. The load resistor must be chosen to yield a dark-current output voltage less than 100 mV.

Given: The detector parameters obtained from Table 17.1 are as follows: NEP = 6.33×10^{-15} W, cathode responsivity $\mathscr{R}_c = 7.9 \times 10^{-2}$ A/W, and cathode dark current $i_c = 5 \times 10^{-16}$ A. In addition, a multiplier gain of $G = 1 \times 10^5$ can be obtained by using a suitable supply voltage.

Basic equations:

$$\mathscr{R}_A = \mathscr{R}_c G \qquad\qquad (17.15)$$

$$T_c = R_L\left(\frac{C_T}{A + 1} + C_L\right) \qquad\qquad (18.28)$$

$$A = \frac{k}{f + a} \qquad\qquad (18.29)$$

Assumptions: The load resistor shunt capacitance is 0.2×10^{-12} F.

Solution: The dark output voltage is given by

$$V_o(\text{dc}) = i_c \times G \times R_L = 5 \times 10^{-16}[\text{A}] \times 1 \times 10^5 \times R_L \text{ [ohms]}$$
$$= 100 \text{ mV}$$

from which

$$R_L = 2 \times 10^9 \text{ ohms}$$

The system responsivity is given by

$$\mathscr{R}(\text{sys}) = \mathscr{R}_A(\text{A/W}) \times R_L[\text{ohms}] = 7.9 \times 10^{-2} \times 1 \times 10^5 \times 2 \times 10^9$$
$$= 1.58 \times 10^{13} \text{ V/W}$$

The input capacitance is 158 pF, which includes 2.0 pF stray and 6 pF FET input capacitances and the 150 pF coaxial cable capacitance. The time constant is

$$T_c = R_L C_T f/k + R_L C_L = 1/2\pi f$$

Using the quadratic equation as in Example 2:

$$\frac{b}{2a} = \frac{kC_L}{2C_T} = \frac{10^6 \times 0.2 \times 10^{-12}}{2 \times 158 \times 10^{-12}} = 632.91$$

$$(b/2a)^2 = 4.006 \times 10^5$$

$$\frac{c}{a} = \frac{k}{2\pi R_L C_T} = \frac{1 \times 10^6}{2\pi \times 2 \times 10^9 \times 158 \times 10^{-12}} = 5.037 \times 10^5$$

$$\left[\left(\frac{b}{2a}\right)^2 - \frac{c}{a}\right]^{1/2} = (4.006 \times 10^5 + 5.037 \times 10^5)^{1/2} = 950.95$$

Thus,

$$f = -632.91 + 950.95 = 318.04 \text{ Hz}$$

where the positive solution is used.

Assuming $C_L = 0$, the solution is

$$f = \left(\frac{k}{2\pi R_L C_T}\right)^{1/2} = \left(\frac{10^6}{2\pi \times 2 \times 10^9 \times 158 \times 10^{-12}}\right)^{1/2} = 710$$

which illustrates the limiting effect of the load resistor time constant as in Example 2. ∎

Examples 2 and 3 illustrate the frequency-response limitations of the load resistor time constant. The following section covers frequency compensation techniques to overcome these limitations.

18.5 FREQUENCY COMPENSATION

Frequency compensation of high-impedance transducers through the use of negative feedback operational preamplifiers has had wide applications. The circuit of the compensated amplifier is given in Fig. 18.11, where the detector is modeled as a high-impedance current source, and the detector, stray, and input capacitances are lumped into C_T. The components R_C and C_C are used to achieve compensation of the load time constant $R_L C_L$.

A node equation is written in the s domain and solved for e_Q as

$$e_Q = \frac{e_o[R_L - R_C(R_LC_Ls + 1)/A]}{R_LR_C(C_C + C_L)s + R_L + R_C} \tag{18.30}$$

Similarly, a node equation is written for point P, using Eq. (18.30), which after some manipulation yields

$$i_s = -e_o\left(\frac{C_Ts}{A} + \frac{R_LC_Ls + 1}{AR_L} + \frac{R_LC_Ls + 1}{R_LR_C(C_C + C_L)s + R_L + R_C}\right.$$
$$\left. + \frac{R_C(R_LC_Ls + 1)^2}{AR_L[R_LR_C(C_C + C_L)s + R_L + R_C]}\right) \tag{18.31}$$

Compensation is achieved when

$$R_CR_L(C_C + C_L) = R_LC_L(R_C + R_L) \tag{18.32}$$

Substitution of Eq. (18.32) into the third term in the large parentheses in Eq. (18.31) yields

$$\frac{R_LC_Ls + 1}{R_LR_C(C_C + C_L)s + R_L + R_C} = \frac{1}{R_C + R_L} \tag{18.33}$$

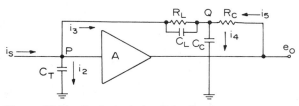

Figure 18.11 Equivalent circuit for the frequency-compensated operational amplifier.

and for the fourth term,

$$\frac{R_C(R_LC_Ls+1)^2}{AR_L[R_LR_C(C_C+C_L)s+R_L+R_C}=\frac{R_C(R_LC_Ls+1)}{AR_L(R_L+R_C)} \quad (18.34)$$

so that Eq. (1831) can be written as

$$i_s=-e_o\left(\frac{C_Ts}{A}+\frac{R_LC_Ls+1}{AR_L}+\frac{1}{R_C+R_L}+\frac{R_C(R_LC_Ls+1)}{AR_L(R_L+R_C)}\right) \quad (18.35)$$

It is desirable to set $R_C \ll R_L$ so the transfer function is dominated by R_L and $C_C \gg C_L$. This also reduces the effect of parasitic capacitances (stray capacitance to ground along R_L and shunt capacitance across R_C). Under these conditions, Eq. (18.35) can be simplified to

$$i_s=-e_o\left(\frac{s(C_T+C_L)}{A}+\frac{1}{AR_L}+\frac{1}{R_L}\right) \quad (18.36)$$

Solving for e_o and letting i_s become a step function as before yields

$$e_o=\frac{A}{C_T+C_L}\left(\frac{1}{s}\right)\left(s+\frac{A+1}{R_L(C_T+C_L)}\right)^{-1} \quad (18.37)$$

which according to Eqs. (18.25) through (18.28) yields a time constant

$$T_c=\frac{R_L(C_T+C_L)}{A+1} \quad (18.38)$$

and all time constants associated with the amplifier are reduced by the gain. For these same conditions (i.e., $R_C \ll R_L$ and $C_C \gg C_L$), Eq. (18.32) can be written

$$R_CC_C=R_LC_L \quad (18.39)$$

as the condition for compensation.

For those frequencies of interest in the operational amplifier time constant of Eq. (18.38), the factor a can be neglected; thus, the time constant is

$$T_c=\frac{R_L(C_T+C_L)f}{k} \quad (18.40)$$

The frequency response is

$$f=\frac{1}{2\pi T_c}=\frac{k}{2\pi R_L(C_T+C_L)f} \quad [\text{Hz}] \quad (18.41)$$

and solving for the upper break frequency f_2,

$$f_2=\left(\frac{k}{2\pi R_L(C_T+C_L)}\right)^{1/2} \quad [\text{Hz}] \quad (18.42)$$

Example 4: Determine the values of R_C and C_C assuming $R_L = 100R_C$, and calculate the frequency response for the InSb detector of Example 1 and a compensated operational amplifier.

Given: In addition to the parameters given in Example 1, the coefficients k and a of Eq. (18.39) (for the 536 amplifier) are 10^6 and 10, respectively.

Basic equations:

$$f_2 = \left(\frac{k}{2\pi R_L (C_T + C_L)} \right)^{1/2} \quad [\text{Hz}] \quad (18.42)$$

$$R_C C_C = R_L C_L \quad (18.38)$$

Assumptions: Ideal electrometer amplifier and no parasitic capacitance.

Solution: It is desirable to have the transfer function dominated by R_L; thus, select $R_L = 100R_C$.
 The value of R_C is

$$R_C = 10^9 / 100 = 10^7 \text{ ohms}$$

and C_C is

$$C_C = \frac{R_L C_L}{R_C} = \frac{1 \times 10^9 \times 0.2 \times 10^{-12}}{1 \times 10^7} = 20 \text{ pF}$$

The input capacitance typically dominates. In this case we have approximately 400 pF input capacitance and 0.2 pF load capacitance; thus, Eq. (18.42) can be written as

$$f_2 = \left(\frac{k}{2\pi R_L C_T} \right)^{1/2} = \left(\frac{10^6}{(2\pi \times 1 \times 10^9 \times 400 \times 10^{-12})} \right)^{1/2} = 630 \text{ Hz}$$

This represents an improvement factor of $630/0.9 = 700$ over the voltage amplifier (see Example 1) and an improvement factor of $630/429 = 1.47$ over the uncompensated transverse impedance amplifier (see Example 2). ∎

Example 5: Determine the values of R_C and C_C assuming $R_L = 100R_C$, and determine the frequency response for the multiplier phototube detector of Example 3 using the compensated operational preamplifier.

Given: See Example 3 for the detector parameters and Example 2 for the amplifier parameters.

Basic equations:

$$f_2 = \left(\frac{k}{2\pi R_L (C_T + C_L)} \right)^{1/2} \quad [\text{Hz}] \quad (18.42)$$

$$R_C C_C = R_L C_L \quad (18.38)$$

Assumptions: Ideal electrometer amplifier and no parasitic capacitance

Solution: The value of R_C is

$$R_C = 2 \times 10^9/100 = 2 \times 10^7 \, \text{ohms}$$

The value of C_C is

$$C_C = \frac{R_L C_L}{R_C} = \frac{2 \times 10^9 \times 0.2 \times 10^{-12}}{2 \times 10^7} = 20 \, \text{pF}$$

The input capacitance of 158 pF dominates; thus,

$$f_2 = \left(\frac{k}{(2\pi R_L C_T)}\right)^{1/2} = \left(\frac{10^6}{(2\pi \times 2 \times 10^9 \times 158 \times 10^{-12})}\right)^{1/2} = 710 \, \text{Hz}$$

This represents an improvement factor of $710/318 = 2.233$ over the uncompensated amplifier. ∎

18.6 NOISE IN THE OPERATIONAL AMPLIFIER

The effect of the input capacitance upon the operation of the compensating amplifier is to cause the loop gain to increase at high frequencies. This is one explanation for frequency compensation, but it also results in amplification of the preamplifier noise. This sets a limit to the bandwidth that can be achieved without introducing excessive noise. The following is an analysis of this effect.

The feedback amplifier can be considered an ideal noiseless amplifier with noise voltage and current sources at the input as shown in Fig. 18.12.

The noise current and noise voltage can be considered to be statistically independent, which means that the total effect is given by the square root of the sum of the squares of each. This permits independent analysis of each effect.

18.6.1 Current Noise

The effect of the noise current can be determined by writing a node equation for point p of Fig. 18.12:

$$i_n = i_1 + i_2 \qquad [\text{A/Hz}^{1/2}] \qquad (18.43)$$

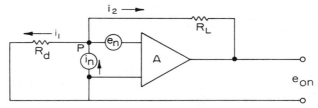

Figure 18.12 Ideal noiseless amplifier with input current and voltage noise sources.

which is

$$i_n = \frac{e_p}{R_d} + \frac{e_p - e_{on}}{R_L} \tag{18.44}$$

this can be written as

$$i_n = \frac{-e_{on}}{R_L}\left[1 + \frac{1}{A}\left(\frac{R_L}{R_d} + 1\right)\right] \tag{18.45}$$

which for large A yields

$$e_{on} = -i_n R_L \qquad [\text{V}/\text{Hz}^{1/2}] \tag{18.46}$$

Typically the input dc bias current is of the order of 10^{-12} A. Although the rms noise i_n arising from the bias current is not likely to be governed by Poisson statistics, it should be considerably less than the average (dc) value. Thus, the input current noise can be neglected for electrometer-type amplifiers.

18.6.2 Voltage Noise

The output noise voltage consists of the product of the open-loop gain and the sum of the noise voltage and a fraction β of the output voltage:

$$e_{on} = (e_n + e_{on}\beta)A \qquad [\text{V}/\text{Hz}^{1/2}] \tag{18.47}$$

The output noise voltage e_{on} is

$$e_{on} = e_n\frac{A}{1 - A\beta} = \frac{e_n}{\beta}\left(\frac{1}{1/A\beta - 1}\right) \tag{18.48}$$

when the product $A\beta$ becomes large,

$$e_{on} = -e_n/\beta \tag{18.49}$$

The value of β for Fig. 18.12 is

$$\beta = \frac{R_d}{R_d + R_L} \tag{18.50}$$

which is a simple voltage divider. Substitution of β into Eq. (18.49) yields

$$e_{on} = -e_n\left(\frac{R_L}{R_d} + 1\right) \tag{18.51}$$

which illustrates that the noise voltage is always amplified by a gain of at least unity.

Equation (18.51) is valid only at frequencies near dc since the effects of capacitance have been neglected. A more detailed analysis can be achieved using the compensated amplifier shown in Fig. 18.13.

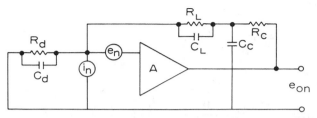

Figure 18.13 Ideal noiseless compensated amplifier with input current and voltage noise sources.

The value of β for the complex network of Fig. 18.13 is

$$\beta = \frac{e_p}{e_{on}} = \frac{R_d}{(R_d R_L C_L + R_L R_d C_T)s + R_L + R_d} \tag{18.52}$$

where $R_c \ll R_f$ and $R_c C_c = R_L C_L$ (the condition for frequency compensation).

Using Eqs. (18.49) and (18.52), the output noise voltage at the frequency f is found to be

$$e_{on}(f) = -e_n[R_L(C_L + C_T)s + 1 + R_L/R_d] \tag{18.53}$$

where $s = j \times 2\pi f$. The break frequency, which is the frequency for which the real and imaginary terms are equal, is given by

$$f_b = \frac{1 + R_L/R_d}{2\pi R_L(C_L + C_T)} \tag{18.54}$$

At frequencies below f_b the noise voltage is given by

$$e_{on}(f) = -e_n\left(1 + \frac{R_L}{R_d}\right) \tag{18.55}$$

and at frequencies above f_b the noise voltage is given by

$$e_{on}(f) = -e_n[j \times 2\pi f R_L(C_L + C_T)] \quad [\text{V/Hz}^{1/2}] \tag{18.56}$$

which increases at the rate of 6 dB per octave.

Example 6:   Find the break frequency for the compensated amplifier and InSb detector of Example 2.

Given: The input noise voltage for the type 536 operational amplifier is 2×10^{-8} V rms/Hz$^{1/2}$ (average between 0.1 Hz and 100 kHz), the total input capacitance is 401 pF, load capacitance is 0.2 pF, load resistor is 1×10^9 ohms, and the detector resistance is 7.8×10^8 ohms.

Basic equation:

$$f_b = \frac{1 + R_L/R_d}{2\pi R_L(C_L + C_T)} \quad [\text{Hz}] \tag{18.54}$$

Assumptions: It is assumed that the amplifier input noise voltage is white (i.e., uniform) between 0.1 Hz and 100 kHz.

Solution: Using Eq. (18.54), the break frequency is found to be

$$f_b = \frac{1 + (1 \times 10^9)/(7.8 \times 10^8)}{2\pi \times 1 \times 10^9 \times 401 \times 10^{-12}} = 0.9 \text{ Hz} \qquad \blacksquare$$

The rms noise voltage is obtained as the modulus of Eq. (18.53) by taking the square root of the sum of squares of the real and imaginary terms. The total mean-square voltage is given by integrating over the frequencies of interest. There are two cases. The first is the narrow-band system, such as a chopped radiometer, where the carrier frequency f_c can be considered constant over the bandwidth Δf. The mean-square voltage is given by

$$\overline{e_{on}^2(f)} = e_n^2 \left\{ \left[2\pi R_L(C_L + C_T) \right]^2 f^2 \Delta f + (1 + R_L/R_d)^2 \Delta f \right\} \quad (18.57)$$

The second is the wide-band system, such as for an interferometer, where the chopping frequency cannot be considered a constant. The mean-square voltage is

$$\overline{e_{on}^2(f)} = e_n^2 \left\{ \left[2\pi R_L(C_L + C_T) \right]^2 \int_{f_1}^{f_2} f^2 \, df + \left(1 + \frac{R_L}{R_d} \right)^2 \int_{f_1}^{f_2} df \right\} \quad (18.58)$$

which is

$$\overline{e_{on}^2(f)} = e_n^2 \left\{ \left[2\pi R_L(C_L + C_T) \right]^2 \frac{f_2^3 - f_1^3}{3} + \left(1 + \frac{R_L}{R_d} \right)^2 (f_2 - f_1) \right\}$$

$$(18.59)$$

where f_1 is the low and f_2 the high edge of the bandpass.

Example 7: Find the output rms noise voltage for the compensated amplifier and InSb detector of Example 2.

Given: The bandwidth is from 100 to 200 Hz, $R_L = 10^9$ ohms, $R_d = 7.8 \times 10^8$ ohms, $C_T = 401$ pF, $e_n = 2 \times 10^{-8}$ V rms (for the type 536 operational amplifier).

Basic equation:

$$\overline{e_{on}^2(f)} = e_n^2 \left\{ \left[2\pi R_L(C_L + C_T) \right]^2 \frac{f_2^3 - f_1^3}{3} + \left(1 + \frac{R_L}{R_d} \right)^2 (f_2 - f_1) \right\}$$

$$(18.59)$$

Assumptions: Ideal amplifier and no parasitic capacitances.

Solution: The rms voltage is obtained as the square root of Eq. (18.59):

$$[2\pi R_L(C_L + C_T)]^2 = (2\pi \times 1 \times 10^9 \times 401 \times 10^{-12})^2 = 6.35$$
$$(f_2^3 - f_1^3)/3 = (200^3 - 100^3)/3 = 2.33 \times 10^6$$
$$(1 + R_L/R_d)^2 = [1 + (1 \times 10^9/7.8 \times 10^8]^2 = 5.21$$
$$f_2 - f_1 = 100$$
$$\overline{e_{on}^2(f)} = (2 \times 10^{-8})^2[6.35 \times 2.33 \times 10^6 + 5.21 \times 100]$$
$$= 5.92 \times 10^{-9}$$
$$e_{on} = 76.9\ \mu\text{V rms} \qquad\blacksquare$$

18.6.3   Total Noise Effects

The output noise for a detector-preamplifier circuit originates in several phenomena that have been discussed in previous sections. They are thermal noise in the detector and in the load resistor, photon noise, and preamplifier noise. Only the preamplifier noise is a function of frequency.

An optimum design is one in which photon noise (or shot noise in a multiplier phototube) dominates. In the absence of photon noise (in low-background systems), the detector thermal-noise limitation is optimum. In many cases, the system is noise-limited by load resistor thermal noise or peramplifier noise. The total output noise for a detector-preamplifier circuit is the square root of the sum of the squares of each term. The designer has to satisfy the sampling theorem requirements first; beyond that, the frequency and load resistor can be varied to obtain a minimum-noise design.

The preamplifier mean-square noise is give by Eqs. (18.57) and (18.59). The load resistor mean-square thermal noise is manifest in the preamplifier output noise as

$$\overline{e_{Ln}^2} = 4kTR_L\,\Delta f \qquad [\text{V}^2] \tag{18.60}$$

The detector thermal noise voltage is manifest in the preamplifier output noise as

$$\overline{e_{dn}^2} = 4kT\Delta f\frac{R_L^2}{R_d} \tag{18.61}$$

The theoretical preamplifier output noise voltage for photon noise input is determined by solving Eq. (16.19) for NEP (for detector cutoff wavelength, temperature of the background, the field of view), and using the detector current responsivity $\mathscr{R}_c$ (A/W) to find the photon noise current [Eq. (6.2)] expressed in terms of current, $i_{pn} = \mathscr{R}_c \times \text{NEP}$. The detector-preamplifier circuit output noise voltage is therefore given by

$$\overline{e_{pn}^2} = i_{pn}^2 R_L^2 \qquad [\text{V}^2] \tag{18.62}$$

The total preamplifier output noise voltage is given by the square root of the sum of the squares of each term[10]

$$e_{nt} = \left(\overline{e_{on}^2} + \overline{e_{dn}^2} + \overline{e_n^2} + \overline{e_{pn}^2}\right)^{1/2} \quad \text{[V rms]} \quad (18.63)$$

Example 8: Select the load resistor and chopping frequency for the InSb detector of Fig. 18.4 that results in the minimum noise equivalent power (NEP).

Given: A compensated preamplifier has an input noise voltage of 1×10^{-8} V rms/Hz$^{1/2}$, the total input capacitance is 393 pF, the required electrical bandwidth is 10 Hz, the minimum chopping frequency is 50 Hz (as determined by the sampling theorem), and the detector current responsivity is $\mathscr{R}_c = 2.0$ A/W. The detector and load resistor are both operated at a reduced temperature of 80 K, and the detector is exposed to a background at 80 K.

Basic equations:

$$\overline{e_{Ln}^2} = 4kTR_L\,\Delta f \quad (18.60)$$

$$\overline{e_{dn}^2} = 4kT\Delta f\frac{R_L^2}{R_d} \quad (18.61)$$

$$\overline{e_n^2} = e_{on}\left\{[2\pi R_L(C_L + C_T)]^2 f^2 \Delta f + \left(1 + \frac{R_L}{R_d}\right)^2 \Delta f\right\} \quad (18.57)$$

$$e_{nt} = \left(\overline{e_{on}^2} + \overline{e_{dn}^2} + \overline{e_{Ln}^2} + \overline{e_{on}^2}\right)^{1/2} \quad (18.63)$$

$$\text{NEP} = e_{nt}/\mathscr{R} \quad \text{[W]} \quad (6.2)$$

$$f_2 = \left(\frac{k}{2\pi R_L(C_T + C_L)}\right)^{1/2} \quad (18.42)$$

Assumptions: Ideal amplifier and no parasitic capacitance.

Solution: The photon noise is negligible in this case because of the low background. The chopping frequency is also fixed at 50 Hz by the sampling theorem and practical filter considerations. Thus, the task consists of selecting the load resistor. The noise terms are e_{dn} (detector thermal noise voltage), e_{Ln} (load resistor noise voltage), e_{on} (preamplifier noise voltage), and e_{nt} (total output noise voltage). The noise terms are tabulated in Table 18.1. ∎

The tabulated data of Example 8 are interpreted as follows: The detector noise voltage increases linearly with R_L; the load-resistor noise voltage increases with the square root of R_L; the preamplifier noise voltage breaks at $R_L = 1 \times 10^7$ ohms and increases linearly with R_L thereafter.

Table 18.1

R_L, ohms	Noise Voltage (rms)				NEP, W	f, Hz
	e_{dn}	e_{Ln}	e_{on}	e_{nt}		
10^5	7.5E-10^a	6.6E-8	3.2E-8	7.4E-8	3.7E-13	6.4E + 4
10^6	7.5E-9	2.1E-7	3.2E-8	2.1E-7	1.1E-13	2.0E + 4
10^7	7.5E-8	6.6E-7	5.1E-8	6.7E-7	3.4E-14	6.4E + 3
10^8	7.5E-7	2.1E-6	3.9E-7	2.3E-6	1.1E-14	2.0E + 3
10^9	7.5E-6	6.6E-6	3.9E-6	1.1E-5	5.4E-15	6.4E + 2
10^{10}	7.5E-5	2.1E-5	3.9E-5	8.7E-5	4.4E-15	2.0E + 2
10^{11}	7.5E-4	6.6E-5	3.9E-4	8.5E-4	4.3E-15	6.4E + 1
10^{12}	7.5E-3	2.1E-4	3.9E-3	8.5E-3	4.2E-15	2.0E + 1
10^{13}	7.5E-2	6.6E-4	3.9E-2	8.5E-2	4.2E-15	6.4E + 0
10^{14}	7.5E-1	2.1E-3	3.9E-1	8.5E-1	4.2E-15	2.0E + 0

^{a} 7.5E-10 is computer notation for 7.5×10^{-10}.

The total noise voltage is dominated by detector thermal noise voltage for values of $R_L > 10^9$ ohms, which is optimum for low-background systems. NEP bottoms out at 4.20×10^{-15} W. The frequency response requirement limits R_L to about 10^{12} ohms; however, large values of R_L limit the dynamic range, since the output noise voltage is getting so large. A compromise choice is $R_L = 10^9$ ohms.

For $R_L = 10^9$ ohms, the frequency response is 640 Hz and NEP = 5.4 $\times 10^{-15}$ W. The dynamic range is given by the ratio of full-scale output (approximately 10 V for ± 15 V supplies) to the noise voltage, which in this case is 11 μV.

The solution given in Example 8 can be extended to include a prediction of the noise equivalent sterance [radiance] by making use of the throughput (the detector area and f-number of the system). This complicates the analysis, since increased detector area not only increases the throughput but also increases the input capacitance and the preamplifier noise. An optimum solution is not obvious. Circuit analyses like that of Example 8 require computer subroutines to permit a search for the optimum load resistor, chopping frequency, and detector size.

EXERCISES

1. Referring to manufacturers' data books:
   (*a*) List input current for typical operational amplifiers that utilize (1) bipolar, (2) JFET, and (3) MOSFET input devices. *Hint:* Survey typical manufacturers' specification data for linear devices.
   (*b*) Comment upon their usefulness as electrometers.
   (*c*) Calculate the virtual ground input resistance (at dc) for a 10^7-ohm load resistor.
   (*d*) Calculate the dc error due to bias current.
   (*e*) Note the input noise voltage.

2. Determine the values of R_C and C_C assuming $R_L = 1000 R_C$ for a high-impedance photon detector and compensated operational amplifier. Given: $R_L = 10^7$ ohms, $C_L = 1$ pF, $C_T = 200$ pF, $k = 10^7$ (gain bandwidth). Find the frequency response, taking into account that the gain is a function of frequency.

3. A compensated preamplifier and high-impedance detector have the following characteristics: Short-circuit input noise $e_n = 1 \times 10^{-8}$ V rms/Hz$^{1/2}$, $C_T + C_L = 100$ pF, load resistor $R_L = 10^9$ ohms, detector resistance $R_d = 10^{14}$ ohms. Find the break frequency where the real and imaginary preamplifier noise terms are equal.

4. Given that the gain bandwidth is 10^6, find the rms noise voltage per $\sqrt{\text{Hz}}$ for the preamplifier output of Exercise 3 (consider the effect of preamplifier noise only) at 100 Hz.

5. Repeat Exercise 4 except obtain the total noise integrated from 100 to 200 Hz for a bandpass application.

6. Given that the operating temperature of the preamplifier is 300 K, find the thermal noise in the band 100 to 200 Hz for $R_L = 10^9$ ohms (Exercise 5). Compare this with the preamplifier noise voltage (Exercise 5).

REFERENCES

1. R. A. Smith et al., *The Detection and Measurement of IR Radiation*, Oxford University Press, London, 1957, p. 95.

2. C. Belove and M. M. Drossman, *Systems and Circuits for Electrical Engineering Technology*, McGraw-Hill, New York, 1976, p. 574.

3. C. L. Wyatt et al., "A Direct Coupled Low Noise Preamplifier for Cryogenically Cooled Photoconductive ir Detectors," *Infrared Physics*, **14**, 165–176 (1974).

4. D. J. Baker and C. L. Wyatt, "Irradiance Linearity Corrections for Multiplier Phototubes," *Applied Optics*, **3**, 89–91 (1964).

5. D. A. Allenden, "A Feedback Electrometer Amplifier," *Electronic Engineering*, **30**, 31 (1958).

6. I. Pelchowitch and J. J. Zaalberg Van Zelst, "A Wideband Electrometer Amplifier," *Rev. Sci. Instrum.*, **23**, 73–75 (1952).

7. J. Praglin and W. A. Nichols, "High Speed Electrometers for Rocket and Satellite Experiments," *Proc. IRE*, **48**, 771–779 (1960).

8. J. J. Brophy, *Basic Electronics for Scientists*, McGraw-Hill, New York, 1977, p. 305.

9. L. A. Ware and G. R. Town, *Electrical Transients*, Macmillan, New York, 1954, p. 25.

10. P. R. Bevington, *Data Reduction and Error Analysis for the Physical Sciences*, McGraw-Hill, New York, 1969, p. 79.

chapter *19*

Calibration and Error Analysis

19.1 INTRODUCTION

This chapter is concerned primarily with the quality of measurements obtained with radiometric measurement systems. The uncertainties associated with radiometric measurements result from several factors. Radiant emission arises from discrete, random processes, and a radiometer is a device that, at best, provides an estimate of the power or photon rate in this process.[1] Also, uncertainties result from the nonideal response properties of the sensor. The extent of these errors depends upon the nature of the target source and the background as well as the sensor characteristics. Finally, uncertainties are associated with the calibration procedures and reference standards.

Accuracy implies absolute standards and possibly traceable standards. The concept of absolute standards has less meaning for radiation measurements than it does for the standard meter, for example. To begin with, radiometric entities are not basic units like length, mass, or time; rather, they are derived entities with typical units of joules per second, photons per second, or joules per second per area per solid angle per wavelength, etc. In practice, the simulated blackbody has been utilized as a radiometric standard because it can, in principle, be reproduced in any laboratory, and radiometric standards can be traced to units of length and absolute temperature. The accuracy of a standard source is known primarily in terms of the accuracy of its temperature.

19.2 CALIBRATION UNCERTAINTY

The calibration of a radiometric measurement system can be regarded as an experiment in which the output is observed in response to a standard source. It is observed that in a series of such experiments each measurement yields different values. An observation of the measurements indicates that the output can be regarded as a random variable x.

We would like to be able to determine the probability of getting any particular value of x in one measurement. To do this we require a knowledge of the distribution of an infinite number of measurements. This collection of an infinite number of measurements is known as the "parent distribution."

In any practical experiment we obtain only a finite data set. These data represent a sample set of the parent population.

The mean of the parent population is, by convention, the best estimate of the parameter x and is given by[2]

$$\mu = \lim_{n \to \infty} \left(\frac{1}{N} \sum_{i=1}^{N} x_i \right) \qquad (19.1)$$

where the mean is obtained as a limit of an infinite number of measurements.

The deviation of any measurement x_i from the mean μ is defined as the difference

$$d_i = x_i - \mu$$

The average deviation is zero by virtue of the definition of the mean. A more appropriate measure of the dispersion or spread of the points about the mean is the variance s^2, which is defined as the limit of the average of the squares of the deviations from the mean μ,

$$s^2 = \lim_{N \to \infty} \left[\frac{1}{N} \sum (x_i - \mu)^2 \right] \qquad (19.2)$$

and the square root of the variance is the standard deviation s.

In any practical experiment we make only a finite set of measurements, say N observations of x_i, where i ranges from 1 to N. We assume that the probability of obtaining any particular measurement x_i is determined by the parent distribution. However, our experimental set of the N observations represents a sample set of parent values, and all calculations must be made from the sample set. The sample set only permits us to estimate the mean and standard deviation.

For a series of N observations, the most probable estimate of the mean μ is the sample mean[3]

$$\bar{x} = \frac{1}{N} \sum x_i \qquad (19.3)$$

and the sample variance is given by

$$\sigma^2 = \frac{1}{N-1} \sum (x_i - \bar{x})^2 \tag{19.4}$$

The question immediately arises as to the number of measurements, N, required to make a realistic estimate of the parent mean. This is given by the standard error

$$\varepsilon = \left(\frac{1}{N(N-1)} \sum (x_i - \bar{x})^2 \right)^{1/2} = \sigma/\sqrt{N} \tag{19.5}$$

which is the standard deviation divided by the square root of the number of observations.

Example 1:   Find the uncertainty in the sample mean.

Given: A series of 10 measurements are taken for which the mean and standard deviation are 5.32 and 0.45, respectively.

Basic equation:

$$\varepsilon = \sigma/\sqrt{N} \tag{19.5}$$

Solution: Equation (19.5) gives the uncertainty in the estimate of the mean:

$$\varepsilon = 0.45/\sqrt{10} = 0.14$$

The relative uncertainty is

$$(0.14/5.32) \times 100 = 2.67\%$$

which implies that the mean is known quite well for 10 measurements. ∎

According to the law of large numbers,[4] the sample mean and variance are better estimators of the parent random process as the number of measurements increases without limit.

As stated above, we would like to know the probability of getting a particular value of x in any one measurement. More specifically, in calibrating a radiometric sensor, we would like to know, based upon certain calibration experiments, what the probable error is.

The error for any one measurement can be described only in terms of a probabilistic statement based upon the mean and standard deviation and by assuming a *normal* or Gaussian probability function. The Gaussian assumption is used by convention. It is justified because (1) it generally gives good results and (2) the central limit theorem[5] assures that the sample mean becomes Gaussian as the number of statistically independent measurements increases without limit, regardless of the probability distribution of the parent process, provided it has a finite mean and variance.

The error for any one measurement is given in terms of the standard deviation σ. The data in Table 19.1, computed from the normal function,[6]

Table 19. THE PROBABILITY THAT THE ERROR WILL NOT EXCEED $\pm\sigma$

Uncertainty	Probability, %
1.0σ	68.26
1.5σ	86.64
2.0σ	95.44
2.5σ	98.76
3.0σ	99.74

indicate that a 95% confidence level is obtained for 2σ. This can be interpreted two ways: (1) the probability is 95% that the value of x for a particular measurement will differ from the mean value by less than $\pm 2\sigma$, and (2) that for a series of measurements of x, some 95% of the values measured should fall within the range of $\bar{x} \pm 2\sigma$.

19.3 PRECISION AND ACCURACY

There are two types of errors associated with measurements: (1) random and (2) systematic. The effect of random errors can often be reduced by repeating the measurement—or by averaging. Such a process increases the *precision* of a measurement. However, the effect of systematic errors cannot be reduced by averaging. Thus, a measurement may be precise but at the same time inaccurate.

The difference between the mean and the "true" value of an entity is a measurement of the *accuracy*. Thus, a measurement may be accurate although imprecise. This is illustrated as follows: Even in the presence of large random errors, an accurate measurement may be obtained by repeating the measurement and averaging many times to get the true value. However, this is not the case with systematic errors even if the random error is zero. In general, an accurate measurement gives truth; a precise measurement is repeatable.

The *precision* of a measurement is a measure of the reproducibility or consistency of measurements made with the same sensor. The *accuracy* of a measurement refers to absolute measurements and implies absolute standards.[7,8] Neither the precision nor the accuracy can be determined exactly but must be estimated from the sample data set.

19.4 SOURCES OF ERROR

Precision and accuracy in radiometric instrumentation are tied into the multivariable functional relationship between the incident flux and the instrument output.[9]

The instantaneous responsivity is a function of the geometrical variables θ and ϕ, field of view; wavelength λ, the spectral response function; time t, instrument time constant; and polarization p, the orientation of the

E vector, as follows:

$$\mathcal{R} = f\left[\mathcal{R}_0\mathcal{R}(\theta,\phi)\mathcal{R}(\lambda)\mathcal{R}(p)\mathcal{R}(t)\right] \qquad (19.6)$$

where $\mathcal{R}_0$ has the units of output per unit flux, $\mathcal{R}(\theta,\phi)$ is the relative spatial response function (the field of view), $\mathcal{R}(\lambda)$ is the relative spectral response function (the spectral bandpass), $\mathcal{R}(p)$ is the polarization response function, and $\mathcal{R}(t)$ is the time domain response function. (There may be other terms for spatial, spectral, and/or polarization scanning functions.) Each of these terms is generally peak-normalized.

Successful calibration of a sensor depends on the interrelationship of the terms in Eq. (19.6). For well-designed instruments, the responsivity terms can be independently evaluated. Actually, they are functionally independent only within certain operational limits, and calibration tests must be devised that stay within these limits. The quality of the data obtained with a sensor is relative to the degree to which the parameters of Eq. (19.6) can be independently evaluated.

For example, when measuring the field of view, or the spectral bandpass, it is essential that the instrument be scanned through θ and ϕ or λ slowly enough that the instrument time constant does not distort the shape of the response function. Also, when measuring the spectral bandpass function, the source position must remain constant; likewise, when measuring the field of view, it is important that the source spectral characteristics remain constant.

The precision of a sensor can be evaluated in terms of the spread or consistency of the measured data set for each term in Eq. (19.6). It is first necessary to determine the functional nature of Eq. (19.6). This relates to operation over a range of values of the variables. All systems are not necessarily linear by design or by nature. Various nonlinear schemes are occasionally used to extend the dynamic range. However, the relative response terms of Eq. (19.6) must be evaluated with linearized data.

The transfer function shows the functional relationship between incident flux and the electric output signal. To determine system linearity, it is necessary to devise a method to change the incident flux over the sensor's entire dynamic range while maintaining the spectral, spatial, and polarization properties of the source constant.

The data are processed by best fit, which minimizes the mean square error.[10] The square root of the mean square error is the standard deviation, which gives the probability that any one particular measurement will yield the output signal predicted by the functional relationship. This is referred to as the linearity uncertainty σ_L.

The response for each point should be averaged sufficiently to eliminate the effect of internally generated noise. The noise depends on the sensor time constant.

A statistical analysis of the "dark noise" is obtained by placing a lighttight cover over the instrument aperture and providing a sufficiently long run to evaluate the mean and standard deviation. The output may contain noise forms that are coherent functions of scan position, microphonics, dc offset, or drift. The mean (or long-term expectation value) will provide a measure of the false signal content of the output. The "offset error" must be subtracted from all subsequent measurements. The variance provides a measure of the error spread due to system noise.

The sensor precision is given by

$$\sigma_p = \left[\sigma_L^2 + \sigma_{\theta,\phi}^2 + \sigma_\lambda^2 + \sigma_p^2 + \sigma_n^2 \right]^{1/2} \tag{19.7}$$

assuming that each parameter in Eq. (19.6) can be independently evaluated[11] and that the uncertainty or spread in the data can be determined by best fit or otherwise estimated.

The nature of the application of radiometric instrumentation determines the need for precision and/or accuracy. For example, data processing algorithms that are based upon ratios of appropriate radiant entities require only that the measurements be precise; in other words, systematic errors cancel out. Absolute measurements require the instrument to be both precise and accurate.

Systematic errors must be assumed zero in a standard source, since if any were known they would be eliminated. The source uncertainty can usually be traced back to the uncertainty of a standard detector or to absolute temperature and emissivity measurements. The source uncertainty is characterized by σ_s, which includes such factors as resettability, repeatability, and stability.

The radiant input to the sensor depends on the geometrical relationship of the source to the sensor. The problem of determining the uncertainty of the geometrical variables can sometimes be avoided when, for example, an extended-area source is used for an extended-area calibration. Assuming that the source is noise-free and exhibits short-term stability, the sensor precision σ_p can be considered independent of the source. Thus the sensor accuracy σ_a is

$$\sigma_a = \left(\sigma_s^2 + \sigma_p^2 \right)^{1/2} \tag{19.8}$$

Typical values of the uncertainties given in Eq. (19.7) depend upon the sensor design, the flux entity of interest, and the calibration procedure. Generally, the spectral parameters are the most difficult to measure and consequently will dominate in estimating sensor precision.

19.5 SOURCE UNCERTAINTY

Blackbody simulators are generally utilized as standard sources. The effective flux for a radiometric calibration can be approximated by Planck's equation (see Sec. 10.1) for the sterance [radiance] in units of W cm^{-2} sr^{-1} as follows:

$$L = \Delta\lambda\, C_1 \lambda^{-5} \left[\exp(C_2/\lambda T) - 1 \right]^{-1} \tag{19.9}$$

where L is the sterance [radiance], W cm^{-2} sr^{-1}

     $\Delta\lambda$ is the radiometer bandwidth, μm

     λ is the band center wavelength, μm

     T is the source absolute temperature, K

     $C_1 = 1.191066 \times 10^4$

     $C_2 = 1.43883 \times 10^4$

The uncertainty in the sterance [radiance] resulting from a temperature uncertainty can be determined by taking the derivative of Eq. (19.9) with respect to temperature:

$$\frac{dL}{dT} = \Delta\lambda\, C_1 \lambda^{-5} \left[\exp\left(\frac{C_2}{\lambda T}\right) - 1\right]^{-2} \exp\left(\frac{C_2}{\lambda T}\right)\frac{C_2}{\lambda T^2} \qquad (19.10)$$

which can be written by passing to the incremental form as

$$\frac{\Delta L}{\Delta T} = \frac{C_1 C_2\, \Delta\lambda}{T^2 \lambda^6}\exp\left(\frac{-C_2}{\lambda T}\right) \qquad (19.11)$$

where $\exp(C_2/\lambda T) \gg 1$ (see Sec. 16.2).

Example 2:   Find the required temperature uncertainty for 1% uncertainty in the sterance [radiance] for a band-radiometer calibration.

Given: The radiometer bandcenter is 1.250 μm, bandwidth 0.3 μm, and source temperature 620 K.

Basic equations:

$$L = \Delta\lambda\, C_1 \lambda^{-5}[\exp(C_2/\lambda T) - 1]^{-1} \qquad (19.9)$$

$$\Delta L/\Delta T = \left(C_1 C_2\, \Delta\lambda/T^2 \lambda^6\right)\exp(-C_2/\lambda T) \qquad (19.11)$$

Solution: First find the sterance [radiance], L:

$$\Delta\lambda\, C_1 \lambda^{-5} = 0.3 \times 1.91066 \times 10^4 \times (1.25)^{-5} = 1.88 \times 10^3$$

$$C_2/\lambda T = 1.43883 \times 10^4/(1.25 \times 620) = 18.6$$

$$\exp(C_2/\lambda T) = 1.16 \times 10^8$$

$$L = 1.88 \times 10^3/(1.16 \times 10^8) = 1.62 \times 10^{-5}$$

Next, find the sensitivity, $\Delta L/\Delta T$:

$$\frac{C_1 C_2}{T^2 \lambda^6} = \frac{1.43883 \times 10^4 \times 1.191066 \times 10^4}{620^2 \times (1.250)^6} = 116.83$$

$$C_2/\lambda T = 1.43883 \times 10^4/(1.25 \times 620) = 18.57$$

$$\exp(-C_2/\lambda T) = 8.65 \times 10^{-9}$$

$$\Delta L/\Delta T = 116.83 \times 0.3 \times 8.65 \times 10^{-9} = 3.30 \times 10^{-7}$$

The source uncertainty ΔL is specified as 1% of the sterance [radiance], L:

$$\Delta L = 0.01 \times 1.62 \times 10^{-5} = 1.62 \times 10^{-7}$$

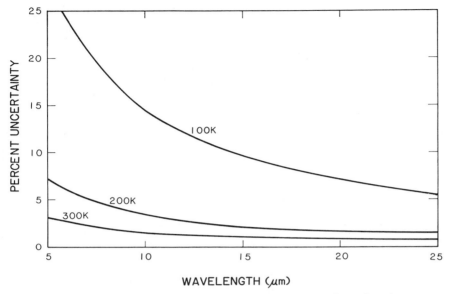

Figure 19.1 Percent uncertainty per unit degree as a function of wavelength.

Thus

$$\Delta T = 1.62 \times 10^{-7}/(3.03 \times 10^{-7}) = 0.532 \text{ K}$$

or

$$\Delta T = (0.532/620) \times 100 = 0.0858\% \quad \blacksquare$$

The solution to Eqs. (19.9) and (19.11) is given in Fig. 19.1, in terms of sterance [radiance] uncertainty for a temperature uncertainty of 1 K. The data show that the uncertainty varies with both wavelength and temperature. Lower temperatures and shorter wavelengths exhibit greater uncertainties.

19.6 NONIDEAL RESPONSE

As indicated in Chap. 5, the effect of nonideal spectral bandpass and field-of-view response functions in a radiometric sensor is to introduce errors into the measurements.

19.6.1 Spectral Response

The objective of the measurement with respect to the spectral domain is to measure the total integrated flux Φ_T in a specified region λ_1 to λ_2:

$$\Phi_T = \int_{\lambda_1}^{\lambda_2} \Phi(\lambda) \, d\lambda \tag{19.12}$$

where $\Phi(\lambda)$ is the spectral distribution of flux over all wavelengths.

The actual sensor output V in response to $\Phi(\lambda)$ is given by the integral of the product of $\Phi(\lambda)$ and the sensor response function $\mathcal{R}_0 \mathcal{R}(\lambda)$:

$$V = \mathcal{R}_0 \int_0^\infty \mathcal{R}(\lambda)\Phi(\lambda)\,d\lambda \tag{19.13}$$

where $\mathcal{R}_0$ is the peak responsivity in units of volts per unit flux and $\mathcal{R}(\lambda)$ is the peak normalized response function.

The measured flux is obtained from Eq. (19.13) by

$$\Phi_m = \frac{V}{\mathcal{R}_0} = \int_0^\infty \mathcal{R}(\lambda)\Phi(\lambda)\,d\lambda \tag{19.14}$$

which, by Eq. (19.12), is equal to Φ_T only if

$$\int_0^\infty \mathcal{R}(\lambda)\Phi(\lambda)\,d\lambda = \int_{\lambda_1}^{\lambda_2}\Phi(\lambda)\,d\lambda \tag{19.15}$$

which occurs for the ideal radiometer when $\mathcal{R}(\lambda) = 1$ over the range λ_1 to λ_2 is zero elsewhere. Such an ideal radiometer will always yield the total flux Φ_T regardless of the distribution of $\Phi(\lambda)$.

The equality in Eq. (19.15) also holds for the case where $\Phi(\lambda) = \Phi_0$, a uniform function of wavelength, in the case of the nonideal radiometer, provided

$$\int_0^\infty \mathcal{R}(\lambda)\,d\lambda = \int_{\lambda_1}^{\lambda_2} d\lambda = \Delta\lambda \tag{19.16}$$

The nonideal sensor yields the same response to a *uniform* source as does the ideal sensor provided they have equivalent area as given by Eq. (19.16).

Generally the spectral bandpass function is not ideal and the flux distribution is nonuniform; consequently, there will be errors unless corrections are made.

Corrections are made as follows: The source spectral distribution, Eq. (19.12), can be expressed as

$$\Phi(\lambda) = k\Phi_r(\lambda) \tag{19.17}$$

where k is the peak flux and has the units of absolute flux and $\Phi_r(\lambda)$ is the peak normalized flux over the region of interest. The sensor output, Eq. (19.13), can then be written as

$$V = \mathcal{R}_0 k \int_0^\infty \Phi_r(\lambda)\mathcal{R}(\lambda)\,d\lambda \tag{19.18}$$

from which k can be obtained from a field measurement

$$k = V\left[\mathcal{R}_0 \int_0^\infty \Phi_r(\lambda)\mathcal{R}(\lambda)\,d\lambda\right]^{-1} \tag{19.19}$$

provided $\Phi_r(\lambda)$ is known from either theory or spectrometric measurements. The quantities $\mathcal{R}_0$ and $\mathcal{R}(\lambda)$ are measured during the calibration.

Then the total flux is given by

$$\Phi_T = k \int_{\lambda_1}^{\lambda_2} \Phi_r(\lambda)\, d\lambda \tag{19.20}$$

where λ_1 and λ_2 can   define any region of interest that satisfies the measurement objectives and for which $\Phi_r(\lambda)$ is known.

The above discussion suggests that the ideal measurement technique is one that simultaneously employs a band radiometer to evaluate k according to Eq. (19.19) and a spectrometer to measure $\Phi_r(\lambda)$.

When corrections cannot be made, it is recommended practice to calculate the measured flux using Eq. (19.14) and report it as "peak normalized."[12]

It is desirable to estimate the extent of the uncertainty in radiometric measurements when $\Phi_r(\lambda)$ is not known and the corrections given above cannot be applied. In this case the spectral response function $\mathcal{R}(\lambda)$ is made as nearly ideal as possible. However, even the so-called "square" interference filters exhibit a ripple structure in the bandpass. An example is given in Fig. 19.2.

The worst-case error occurs when a single-line spectrum is located at an arbitrary wavelength between λ_1 and λ_2. The correct flux value is given for a single line only when it occurs at the wavelength λ_0 for which $\mathcal{R}(\lambda) = 1$. Examination of Fig. 19.2 shows that the output will vary between 86 and 100% of the peak response depending upon the exact location of the line. The worst-case uncertainty corresponds to the peak-to-peak variations in the spectral responsivity function.

The probable error can be reduced provided that $\mathcal{R}(\lambda)$ in Eq. (19.13) is normalized at the *average* response in the region λ_1 to λ_2 rather than at the peak.

It is important to realize that very large measurement errors can occur when the major features in $\Phi(\lambda)$ lie outside the range λ_1 to λ_2 or when intense out-of-band sources lie within the wings of $\mathcal{R}(\lambda)$ (outside of the region λ_1 to λ_2).

19.6.2 Spatial Response

The above discussion of errors associated with the relative spectral response function also applies to the spatial domain. The mathematics are similar except that the single dimension λ, for wavelength, must be replaced in Eqs. (19.12) through (19.20) with the spatial parameters θ and ϕ and the single integrals must be replaced with double integrals.

It is significantly easier to design optical systems for which the spatial response function approaches the ideal (as compared to the spectral domain), in which $\mathcal{R}(\theta, \phi)$ is very nearly unity over the field of view and for which the off-axis rejection is very high, although such designs are costly to implement.

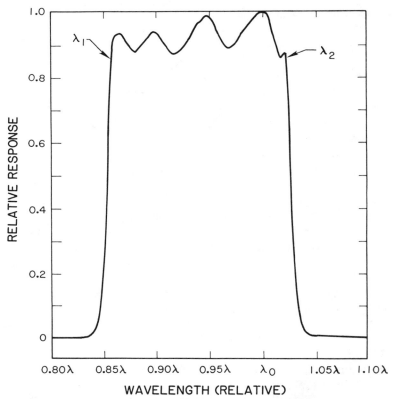

Figure 19.2   Spectral response function of a typical "square" bandpass system showing in-band ripples as a source of radiometric uncertainty.

The uncertainty can be estimated in the case of the spatial response function $\mathcal{R}(\theta, \phi)$ as for the spectral domain given above, by examining the variations in the $\mathcal{R}(\theta, \phi)$ response function, and the probable error can be reduced by normalizing to the average response within the field of view rather than to the peak response.

EXERCISES

1. A convenient computational method for the sample variance is given by

$$\sigma^2 = \frac{1}{N-1} \left[\sum_{i=1}^{N} x_i^2 - \frac{1}{N} \left(\sum_{j=1}^{N} x_i \right)^2 \right]$$

Prove the above relationship given Eq. (19.4).
2. Find the mean for the following sample set: 25.4, 21.3, 22.7, 19.8, 21.5, 25.4, 18.9, 26.4, 22.1, 24.8, 20.4.
3. Find the standard deviation for the data set of Exercise 2.

4. Find the relative (percent) uncertainty in the estimate of the mean for the data set of Exercise 2.
5. Find the sterance [radiance] relative (%) uncertainty for a temperature uncertainty of ± 1 K for the following:
   (a) $T = 1200$ K, $\lambda = 2.5\ \mu$m, $\Delta\lambda = 1\ \mu$m
   (b) $T = 1200$ K, $\lambda = 0.5\ \mu$m, $\Delta\lambda = 1\ \mu$m
6. Find the relative (%) source radiant uncertainty for $T = 300$ K at 3.5 μm given that the temperature uncertainty is ± 1 K and $\Delta\lambda = 1\ \mu$m.
7. Find the required source temperature uncertainty for $T = 1000$ K at 3.5 μm given that the radiant uncertainty must be $\pm 10\%$.

REFERENCES

1. E. J. Kelley et al., "The Sensitivity of Radiometric Measurements," *J. Soc. Ind. Appl. Math*, **11**, 235–257 (1963).
2. P. R. Bevington, *Data Reduction and Error Analysis for the Physical Sciences*, McGraw-Hill, New York, 1969, p. 12.
3. Ibid., p. 18.
4. W. B. Davenport, Jr., and W. L. Root, *An Introduction to the Theory of Random Signals and Noise*, McGraw-Hill, New York, 1958, p. 79.
5. Ibid., p. 81.
6. M. R. Spiegel, *Mathematical Handbook of Formulas and Tables*, (*Schaum's Outline Series*), McGraw-Hill, New York, 1968, p. 257.
7. P. R. Bevington, *Data Reduction and Error Analysis for the Physical Sciences*, McGraw-Hill, New York, 1969, p. 3.
8. F. E. Nicodemus and G. J. Zissis, *Report of BAMIRAC—Methods of Radiometric Calibration*, ARPA Contract No. SD-91, Rep. No. 4613-20-R (DDC No. AD-289, 375), p. 2. Univ. of Michigan Infrared Lab., Ann Arbor, MI, 1969.
9. C. L. Wyatt, *Radiometric Calibration: Theory and Methods*, Academic Press, New York, 1978, p. 85.
10. Ibid., p. 91.
11. P. R. Bevington, *Data Reduction and Error Analysis for the Physical Sciences*, McGraw-Hill, New York, 1969, p. 99.
12. F. E. Nicodemus, "Normalization in Radiometry," *Applied Optics*, **12**, 2960–2973 (1973).

three

APPENDIXES

appendix A

SI Base Units*

Entity	Term	Symbol
Length	meter	m
Mass	kilogram	kg
Time	second	s
Electric current	ampere	A
Thermodynamic temperature	kelvin	K
Luminous intensity	candela	cd

Of the SI base units, the one of particular interest in photometry is the candela, defined as follows: The candela is the luminous pointance [intensity], in the perpendicular direction, of a surface of 1/600,000 square meter of a blackbody at the temperature of freezing platinum under a pressure of 101,325 newtons per square meter.[†] 1 candela = 1 lumen per steradian.

[†]*International Lighting Vocabulary*, 3rd ed. Publ. CIE No. 17(E-1.1), common to the CIE and IEC, International Electrotechnical Commission (IEC), International Commission on Illumination (CIE), Bur CIE, Paris, 1970.

*E. A. Mechtly, "The International System of Units," Rep. NASA SP-7012, pp. 4, 5. Office of Technical Utilization NASA, Washington, DC, 1969.

appendix B

SI Prefixes

Factor	Prefix	Symbol	Factor	Prefix	Symbol
10^{12}	tera	T	10^{-2}	centi	c
10^{9}	giga	G	10^{-3}	milli	m
10^{6}	mega	M	10^{-6}	micro	μ
10^{3}	kilo	k	10^{-9}	nano	n
10^{2}	hecto	h	10^{-12}	pico	p
10^{1}	deka	da	10^{-15}	femto	f
10^{-1}	deci	d	10^{-18}	atto	a

The terms and symbols listed in the tabulation are used, in combination with the terms and symbols, respectively, of the SI units, as prefixes to form decimal multiples and submultiples of those units.*

*E. A. Mechtly, "The International System of Units," Rep. NASA SP-7012, p. 3. Office of Technical Utilization, NASA, Washington, DC, 1969.

appendix C

Atomic Constants*

*E. A. Mechtly, "The International System of Units," Rep. NASA SP-7012, pp. 7, 8. Office of Technical Utilization, NASA, Washington, DC, 1969.

Entity	Symbol	Value	Error, ppm	Prefix	Unit
Speed of light in vacuum	c	2.9979250	000.33	$\times 10^8$	m s^{-1}
Gravitational constant	G	6.6732	460.00	10^{-11}	N m^2 kg^{-2}
Avogadro constant	N_A	6.022169	006.60	10^{26}	kmole^{-1}
Boltzmann constant	k	1.380622	043.00	10^{-23}	J K^{-1}
Gas constant	R	8.31434	042.00	10^3	J kmole^{-1} K^{-1}
Volume of ideal gas (standard conditions)	V_0	2.24136		10^1	m^3 kmole^{-1}
Faraday constant	F	9.648670	005.50	10^7	C kmole^{-1}
Unified atomic mass unit	u	1.660531	006.60	10^{-27}	kg
Planck constant	h	6.626196	007.60	10^{-34}	J s
	$h/2\pi$	1.0545919	007.60	10^{-34}	J s
Electron charge	e	1.6021917	004.40	10^{-19}	C
Electron rest mass	m_e	9.109558	006.00	10^{-31}	kg
		5.485930	006.20	10^{-4}	u
Proton rest mass	m_p	1.672614	006.60	10^{-27}	kg
		1.00727661	000.08	—	u
Neutron rest mass	m_n	1.674920	006.60	10^{-27}	kg
		1.00866520	000.10	—	u
Electron charge-to-mass ratio	e/m_e	1.7588028	003.10	10^{11}	C kg^{-1}
Stefan-Boltzmann constant	σ	5.66961	170.00	10^{-8}	W m^{-2} K^{-4}
First radiation constant	$8\pi hc$	4.992579	007.60	10^{-24}	J m
Second radiation constant	hc/k	1.438833	043.00	10^{-2}	m K
Ryberg constant	R_∞	1.09737312	000.10	10^7	m^{-1}
Fine structure constant	α	7.297351	001.50	10^{-3}	
	α^{-1}	1.3703602	001.50	10^2	
Bohr radius	a_0	5.2917715	001.50	10^{-11}	m
Magnetic flux quantum	Φ_0	2.0678538	003.30	10^{-15}	Wb
Quantum of circulation	$h/2m_e$	3.636947	003.10	10^{-4}	J s kg^{-1}
	h/m_e	7.273894	003.10	10^{-4}	J s kg^{-1}

The Elements and Their Chemical Symbols

Name	Symbol	Name	Symbol	Name	Symbol
Actinium	Ac	Krypton	Kr	Technetium	Tc
Aluminum	Al	Lanthanum	La	Tellurium	Te
Americium	Am	Lawrencium	Lr	Terbium	Tb
Antimony	Sb	Lead	Pb	Thallium	Tl
Argon	Ar	Lithium	Li	Thorium	Th
Arsenic	As	Lutetium	Lu	Thulium	Tm
Astatine	At	Magnesium	Ms	Tin	Sn
Barium	Ba	Manganese	Mn	Titanium	Ti
Berkelium	Bk	Mendelevium	Md	Tungsten	W
Beryllium	Be	Mercury	Hg	Uranium	U
Bismuth	Bi	Molybdenum	Mo	Vanadium	V
Boron	B	Neodymium	Mo	Xenon	Xe
Bromine	Br	Neon	Ne	Ytterbium	Yb
Cadmium	Cd	Neptunium	Np	Yttrium	Y
Calcium	Ca	Nickel	Ni	Zinc	Zn
Californium	Cf	Niobium	Nb	Zirconium	Zr
Carbon	C	Nitrogen	N		
Cerium	Ce	Nobelium	No		
Cesium	Cs	Osmium	Os		
Chlorine	Cl	Oxygen	O		
Chromium	Cr	Palladium	Pb		
Cobalt	Co	Phosphorus	P		
Copper	Cu	Platinum	Pt		
Curium	Cm	Plutonium	Pu		
Dysprosium	Dy	Polonium	Po		
Einsteinium	Es	Potassium	K		
Erbium	Er	Praseodymium	Pr		
Europium	Eu	Promethium	Pm		
Fermium	Fm	Protactinium	Pa		
Fluorine	F	Radium	Ra		
Francium	Fr	Radon	Rn		
Gadolinium	Gd	Rhenium	Re		
Gallium	Ga	Rhodium	Rh		
Germanium	Ge	Rubidium	Rb		
Gold	Au	Ruthenium	Ru		
Hafnium	Hf	Samarium	Sm		
Helium	He	Scandium	Sc		
Holmium	Ho	Selenium	Se		
Hydrogen	H	Silicon	Si		
Indium	In	Silver	Ag		
Iodine	I	Sodium	Na		
Iridium	Ir	Strontium	Sr		
Iron	Fe	Sulfur	S		
		Tantalum	Ta		

Glossary

Symbol	Term	Units*
A	Area	cm^2
A_c	Collector area	cm^2
A_d	Detector area	cm^2
A_{fs}	Field stop area	cm^2
A_p	Projected area	cm^2
A_{ps}	Projected source area	cm^2
A_s	Source area	cm^2
C	Constant	—
C	Capacitance	F
C_C	Compensation capacitance	F
C_d	Detector capacitance	F
C_L	Feedback capacitance	F
C_T	Thermal capacitance	J/K
C_T	Total capacitance	F
c	Velocity of light	cm/s
D, d	Diameter	cm
D^*	D star (detectivity)	$cm\, Hz^{1/2}/W$
E	Areance [irradiance]	ϕ/cm^2
$\mathbf{E}$	Electric field vector	—
E_w	Energy of surface work function	J
e	Charge on an electron	C
e	(subscript) Energy	
e_{jn}	Detector noise voltage	V
e_{Ln}	Load resistor noise voltage	V
e_n	Noise voltage	V
e_{nt}	Total noise voltage	—

307

Symbol	Term	Units*
e_{on}	Preamplifier output noise voltage	V
F	f number	—
F	Configuration factor	—
F	Primary focal point	—
F'	Secondary focal point	—
f	Electrical frequency	Hz
f	Primary focal length	cm
f_2	Low-pass filter cutoff frequency	Hz
f'	Secondary focal length	cm
f_c	Chopping frequency	Hz
f_m	Maximum frequency	Hz
f_s	Sample rate	Hz
G	Gain	—
G_T	Thermal conductance	W/K
H	Magnetic field vector	—
h	Planck's constant	Js
I	Pointance [intensity]	ϕ/sr
I_b	Bias current	A
i	Image distance	cm
k	Multiplier noise factor	—
k	Boltzmann's constant	J/K
$\hat{k}$	Unit vector	—
L	Sterance [radiance]	$\phi\ \text{cm}^{-2}\ \text{sr}^{-1}$
L_s	Source sterance [radiance]	$\phi\ \text{cm}^{-2}\ \text{sr}^{-1}$
L_{eff}	Effective sterance [radiance]	$\phi\ \text{cm}^{-2}\ \text{sr}^{-1}$
M	Areance [exitance]	ϕ/cm^2
MTF	Modulation transfer function	—
m	Multiplier stage gain	—
m	Mass	g
N	Number of resolution elements	—
NEF	Noise equivalent flux	ϕ
NEFD	Noise equivalent flux density	W/cm^2
NEI	Noise equivalent input	W
NEP	Noise equivalent power	W
NES	Noise equivalent sterance	$\text{W cm}^{-2}\ \text{sr}^{-1}$
NESS	Noise equivalent spectral sterance	$\text{W cm}^{-2}\ \text{sr}^{-1}\ \mu\text{m}^{-1}$
n	Index of refraction	—
n	Number of electrons	—
o	Object distance	—
P	Period	s
p	(Subscript) Photon	—
R	Electrical resistance	ohms
R_b	Bias resistance	ohms
R_C	Compensation resistance	ohms
R_d	Detector resistance	ohms
R_f	Feedback resistance	ohms
R_L	Load resistance	ohms
R_T	Thermal resistance	K/W
$\mathscr{R}$	Responsivity	output/ϕ
$\mathscr{R}_c$	Cathode responsivity	A/W

Symbol	Term	Units*
$\mathscr{R}_c$	Current responsivity	A/W
$\mathscr{R}(\theta, \phi)$	Relative spatial responsivity	—
$\mathscr{R}(\lambda)$	Relative spectral responsivity	
r	Radius	cm
$\hat{r}$	Unit vector	—
S	Scan rate	scan/s
s, x, y, z	Distance	cm
T	Temperature	K
T_c	Time constant	s
T_d	Dwell time (integration time)	s
T_r	Rise time	s
T_s	Sample time	s
T_{tr}	Transit time	s
t	Time	s
Δt	Transit time spread	s
U	Energy	J
V	Volume	cm^3
V_b	Bias voltage	V
V_n	Noise voltage	V
V_s	Signal voltage	V
v	Velocity	cm/s, m/s
v	(Subscript) Visible	—
α	Absorptance	—
α	Baffle angle	deg (°)
α	Temperature coefficient of resistance	K^{-1}
β	Chopping factor	—
θ	Polar angle	deg (°)
Υ	Throughput	cm^2 sr
ε	Standard error	—
ε	Emissivity	—
λ	Wavelength	m, μm, nm, A
ν	Optical frequency	Hz
$\bar{\nu}$	Wave number	cm^{-1}
ρ	Reflectance	—
ρ_v	Volume emission rate	$q\ s^{-1}\ cm^{-3}$
σ	Stefan-Boltzmann constant	$W\ m^{-2}\ K^{-4}$
σ_L	Linearity uncertainty	—
σ_n	System noise uncertainty	—
σ_P	Polarization uncertainty	—
σ_s	Source uncertainty	—
$\sigma_{\theta, \phi}$	Field of view uncertainty	—
σ_λ	Spectral uncertainty	—
τ	Sample time	s
τ	Transmissivity	—
τ_e	Optical efficiency	—
τ_p	Path transmittance	—
Φ	Flux (general term)	ϕ
Φ_d	Flux incident upon a detector	ϕ
Φ_e	Radiant flux (energy rate)	W
Φ_i	Incident flux	ϕ

Symbol	Term	Units*
Φ_m	Measured flux	ϕ
Φ_p	Photon flux (quanta rate)	q/s
Φ_s	Source flux	ϕ
Φ_T	Total flux	ϕ
Φ_v	Luminous flux (visible)	lm
$\Phi(\lambda)$	Spectral flux	ϕ/λ
ϕ	Azimuthal angle	deg (°)
Ω	Projected solid angle	sr
Ω_c	Projected solid angle with vertex at collector	sr
Ω_F	Projected solid angle of f-number cone	sr
Ω_s	Projected solid angle with vertex at source	sr
ω	Solid angle	sr
ω_c	Solid angle with vertex at collector	sr
ω_s	Solid angle with vertex at source	sr

*ϕ is a generalized symbol for unit of flux. q = quantum.

INDEX

A

Aberrations, 172, 186
Absorptance, 167–169
  definition of, 50
Accuracy, 286, 289
  in amplitude, 90
Active systems, 5
Aggregate, 50
Air glow, 56–58, 121
Airy disk, 71, 194
Aliasing, 93–95
Angle
  definition of, 34
  see Field of view
  see Projected solid angle
  see Solid angle
APART, 218
Apertures, 182–186
Apostilb, 46
Apparent pointance, 50
Areance
  definition of, 39, 43, 45

B

Background limited infrared photocon-
    ductor (BLIP), 236
Baffling in optical systems, 208–225
  knife-edge, 212

nth-order, 209–212
  primary baffle angle, 213, 222
  sunshade baffle, 209
Beam, 36
Bidirection reflectance distribution
    function, 218–220
Blackbody radiation, 143–154
Blur, 71, 194
Brewster's angle, *see* Polarization
Bridge factor, 264
Brightness, *see* Sterance

C

Calibration
  of system, 12, 286–296
Candela, *see* Luminous pointance
Chamberlain, 57
Chopping factor, 69
Coherence of **E** vector, 5
Coherent rectification, 104–106
Configuration factors
  definition of, 37
Conjugate points, 175

D

D^*, 128–131
  definition of, 81